INTRODUCTION TO FIRE SCIENCE

Second Edition

GLENCOE FIRE SCIENCE SERIES

About the Authors

Loren S. Bush, a consulting fire protection engineer, served on the Board of Directors of the National Fire Protection Association for twenty-five years, the last seven years as chairman. Upon his retirement from the board in 1972, he was elected an honorary life member of the association. He was Chief Engineer of the Board of Fire Underwriters of the Pacific and its successor organization, the Pacific Fire Rating Bureau. A graduate of Stanford University, he has frequently lectured to fire science groups and has written several articles for magazines on fire science subjects.

James H. McLaughlin is Assistant Fire Chief and Fire Marshal for the Fire Department of Palo Alto, California, where he was a training officer, fire captain, and fire fighter. He has served as an officer of several California fire service organizations and has taught courses in fire science at San Jose City College. He is currently president of the California Fire Chiefs Association, Northern Division, Fire Prevention Officers Section.

2 Historical and Scientific Background

Archeologists tell us that from earliest days man has known fire. Probably the first introduction to fire came from forest fires started by lightning or volcanoes; so man first met fire as an enemy, something to be stopped or avoided. It must have taken a long time to learn that this terrible thing could be useful. But eventually, man learned to use fire for warmth, for protection against wild animals, and for cooking food.

In time, man not only learned various uses of fire, but also how to start one by rubbing two sticks together. A similar device still used by some primitive tribes is the fire drill. This device consists of a bow with a loose string that is looped around a hardwood stick or drill, and a flat piece of softer wood. Moving the bow in a sawing motion turns the drill fast enough to generate enough frictional heat to ignite tinder. When early man began to chip flint to make tools, he discovered sparks could be struck from stone, and these sparks, in turn, would ignite tinder for a fire. This method of making fire is similar to the flint and steel used in comparatively modern times. The match that lets us start a fire easily was not invented until 1827.

Early Fire Protection

Vents for smoke in the caves of early humans show that from the beginning of attempts to use fire, they had to know something about fire behavior. Primitive humans probably did not do anything to prevent forest fires but, like other animals, concentrated on self-preservation by escape. In historical times, however, organization for fire protection was one of the marks of civilization and, in fact, predated local government.

The need for organized fire protection, under local control, was a strong influence in developing local government as a counterforce against absolute control from a central state. Early forms of municipal government seem to have used the service areas of the fire department as the boundaries of local government jurisdiction.

The "Vigiles," one of the first fire protection organizations found in history, was organized by the Roman Emperor Augustus about the time of Christ. It was a group of the Emperor's slaves who acted as night watchmen with some of the functions of a fire department as we know them today. Even after the fall of the Roman Empire, the Vigiles had proved to be so valuable that they were maintained for many years to protect Roman cities. These Vigile organizations are now generally considered to be the forerunners of our modern fire departments. Probably because of destruction by fire, not enough historical records survive to give us any real idea of what other kinds of organized fire protection existed.

Roots of Modern Organizations

In 1648, a public fire organization was founded in New Amsterdam, the Dutch colony that became New York City. Although the function of this organization was primarily fire prevention (little or no effort was directed toward fire extinguishment), it is usually recognized as the first modern fire department.

In England, after the great fire of London in 1666, fire insurance companies established fire brigades for the benefit of their policyholders, a practice that spread to America. When a company issued a policy insuring a building against fire loss, it provided the owner with a cast-iron plaque bearing the symbol of the insurance company. These were known as "fire marks" and were attached to the building in a prominent place. Following a fire alarm, in those days given by people shouting, the fire brigades would all respond. But, if the first brigade to arrive at the fire did not see its own company's "fire mark" (see Figure 2-1), it would leave and let the fire burn until the proper brigade arrived and tried to save the building. Although limited in purpose compared to our modern organizations, these company brigades operated for nearly two hundred years and developed many techniques of fire fighting still in use today.

About fifty years after the New Amsterdam beginning, Boston set up a more complete public fire fighting organization. This consisted of a group paid from city funds and supplied with equipment for extinguishing fires. Later, volunteer companies known as "mutual fire societies" were organized in Boston to supplement the work of the fire department. The function of these groups was to carry out furniture and other personal effects during the fire; in other words, to do salvage work.

FIGURE 2-1. Old company fire marks. Courtesy of the Palo Alto Fire Department. Photo by Jim McGee

After seeing the mutual fire societies operate in Boston, Benjamin Franklin (in 1736) organized volunteer fire companies to do both fire fighting and salvage work. He is generally credited with being the originator of the volunteer fire departments that have played such an important role in the development of fire protection in this country.

By the mid-1800s, it was generally conceded that fire department operation should be a governmental function. The insurance companies began to eliminate their brigades, although as a later development they sponsored salvage companies. One of the old-timers who started in a volunteer department during the late 1800s explained this development:

"We had two methods of attack—we laid out the hose lines and, if we couldn't wash the building off the lot, we went back and got the axes and, by golly, we brought it down."

Although this is an exaggeration, it was quite common for water damage to exceed the loss by fire. For this reason, fire insurance firms maintained salvage companies in many large cities. These auxiliary organizations would respond to all fire alarms and would protect buildings and contents against water damage. The first such company was organized in 1861, but they were gradually discontinued as fire departments assumed salvage responsibility.

Development of Modern Fire Protection

During the nineteenth and twentieth centuries, the history of fire protection development in the United States involved two major components: (1) changes in equipment, and (2) changes in organization.

Evolution of Fire Equipment

Historians tell us that most of the fires in early American cities were caused by faulty chimneys. Cracks or defects allowed fire or hot gases to reach surrounding structural members, and sparks or burning embers ejected from the stack fell on thatched or shingled roofs of nearby dwellings. The usual method of extinguishing these fires was a bucket brigade, using leather buckets from the nearest water supply. The force pump was first developed by the Roman Vigiles, but the art of making it was lost after the fall of the Roman Empire. The principle was rediscovered in the eighteenth century and developed into the hand-pumper in the early nineteenth century. Some of the largest size could pump about 1140 liters* (300 gallons) per minute with about 50 men on the handles, but even a crew of 50 had strength enough for only a short time. Riveted leather hose was the first type used to carry water from the pump to the fire. Not until the late 1800s did rubber-lined cotton hose become available.

About the middle of the nineteenth century, public water systems began to be used. They provided fire fighters with a handy source of water. The wooden mains could be tapped by boring a hole in them. Each pumper carried a short piece of pipe that could be pushed into the hole and would deliver water to the pump. When the department no longer needed water, the short pipe was pulled out and replaced with a wooden plug. Hence, the name "fire plug," still used to refer to modern fire hydrants. But, as buildings began to grow taller and bigger, effective fire fighting called for more water and greater range. Fortunately, the steam engine was soon developed, allowing fire departments to adopt steam pumps. These steam pumpers, splendid creations with large, spoked wheels and shiny metal work, were horse drawn, usually in a three-abreast hitch, creating a spectacular sight as they raced to the fire. (See Figure 2-2.)

*For a discussion of the metric system, see Appendix A.

FIGURE 2-2. Steam pumper drawn by three galloping horses races to a fire in suburban Los Angeles. Courtesy of the Los Angeles City Fire Department.

About the same time, chemical engines appeared on the fire scene. The first models were hand-drawn and consisted of a large tank, 150 to 230 liters (40 to 60 gallons), built to withstand high pressures and mounted between two large wheels. The tank was filled with a mixture of water and baking soda (sodium bicarbonate). A separate container within the tank held sulfuric acid and, by turning a wheel on the outside of the tank, the acid could be dumped into the soda water. Chemical reaction formed a gas under high pressure within the tank; this gas forced water through a small attached hose. The device was quite effective on small fires but, of course, it was limited by the small amount of water within the tank. Chemical tanks, first mounted on wagons and later on trucks, supplied the small-stream demands of fire departments well into the twentieth century.

The Pace Quickens

Perhaps it was a series of conflagrations at the turn of the century, such as the great Chicago and San Francisco fires, that awakened the public to the need for good fire protection; or perhaps this increased concern was simply a natural development of the industrial age. At any rate, fire protection made important steps forward during the 1900s and its evolution has seemed to accelerate with the years. One of many dramatic changes was the retirement of horses in favor of automotive equipment. As the automobile developed reliability and power, it took over more of the work load of the fire department. For example, a separate hose carrier was no longer necessary because the hose could now be carried on the pumper. Lightweight water tanks, with small motorized pumps that drew from the tank and discharged through a small hose line, replaced chemical engines.

Many other developments in equipment have had a marked effect on fire department operations. Such advances as radio for communication, the use of gas masks and other breathing appliances, and other major measures will be discussed in detail later.

Changes in Organization

At the beginning of the nineteenth century, most fire departments consisted of independent volunteer companies, with minimal cooperation between them. Reports show that some insurance companies offered rewards in the form of a donation to the first company at the scene of a fire. This created such competition that companies frequently spent more time battling each other than fighting the fire. Obviously, this was unprofitable for both the fire fighters and the insurance companies. As a result, companies began to specialize; one as a hose company, another as an engine company or a ladder company; and all came to work as one team.

The transformation of fire departments from volunteer to paid has been slow and is still going on. In fact, some fairly large cities and many of our smaller communities are still effectively served by volunteers. One early change followed the introduction of horses into the service. Many volunteers who had looked upon the fire service as an athletic club now lost interest, so that department members soon had to be paid. Also, more skill was required to operate the steam pumpers, making trained workers essential for the job. Some cities met this problem by hiring full-time drivers and stokers, while continuing to depend upon local citizens who were paid per call to complete the fire fighting force.

Today's semimilitary type of fire organization had its origin in New York City after the Civil War. The department was organized into battalions and divisions, and personnel were ranked as in the Union Army. This line of command is still in general use today. (See Chapter 4.)

Working conditions for fire fighters improved greatly between 1860 and 1920, and still more after 1930. Up until the 1920s, the common workweek for a fire fighter was six 24-hour days. Quarters were in the "fire barn" and the pay was even lower than for a common laborer. Naturally, this did not attract high quality personnel. A person who took a job in the fire department usually did so with the idea of staying only until something better was available. But this did not apply to all. There were some who, after being in the department, could see possibilities of making the fire service both a highly skilled and a well-rewarded occupation. To them the fire fighters of today owe a debt of gratitude for the past 150 years of improvement in department standards and welfare.

Early Steps Toward Formal Education in Fire Science

Many examples of the historic improvements in fire departments could be cited, but an important point is that the fire service started improving itself by means of self-education and training. Later, a few cities established fire colleges and states began to assist through their departments of education and universities. All this resulted in educational resources for even small departments. Early in the century, too, a surge of interest was manifest in fire prevention. Within departments, special bureaus were organized to carry out this phase of fire protection. The study of fire prevention developed just as did the study of extinguishment.

The result of the ambition and effort of the old timers and their successors was to lift the service to high levels of proficiency, respectability, reasonable working conditions, and compensation. Probably the greatest benefit of all has been the change in the type of person who has been attracted to the fire service: one who has a strong desire to make fire protection a *career* and who is willing to study and work not only to advance personally, but also to raise the service to a still higher standard.

Fire, the Destroyer

Death and Injury by Fire

History reveals that human beings have felt a need for fire protection from the beginning, and as civilization advanced they became even more aware of this need. Yet the current principal source of fire information, the National Fire Protection Association (NFPA), estimates that in recent years the annual death rate due to fire in the United States has been about 56 per million of population. This makes fire third on the list of causes of accidental deaths, surpassed only by traffic accidents and falls. Injuries from fire are estimated to be at least ten times the number of deaths each year. A total of 123,000 injuries for the year 1974 has been estimated by the NFPA. Fortunately, the total of deaths by fire has been declining since 1970 and because of improvements in medical techniques, the chances of recovery from serious fire injuries have greatly improved.

To provide more data to help prevent future deaths and injuries by fire, by far the most serious problem in fire science, it is important to know who the casualties are and where the fires occur. A study made by the NFPA shows that the very young (under 5 years) and the old (over 75 years) suffered the greatest percentage of fire deaths in recent years compared to the total population.

According to a report made by the International Association of Fire-fighters (IAFF), in 1976, 70 out of every 100,000 full-time fire fighters lost their lives in line of duty.

Where did loss of life fires occur? More than 93 percent occurred in residential properties: 63.5 percent in one- and two-family dwellings, 22 percent in apartments, 3 percent in hotels, and 4 percent in other residential properties. Only 3 percent occurred in industrial properties, and 4.5 percent occurred in other areas.

Like conflagrations, large loss-of-life fires seem to have been more prevalent in the early part of this century. The largest one of all occurred in 1904 when the excursion steamer General Slocum burned on the East River in New York City with a loss of 1030 lives. Just six months before, the Iroquois Theater in Chicago burned and took the lives of 602 persons, due mainly to panic caused by inadequate exits. The third most disastrous fire, causing a loss of life of 492 persons, happened in Boston when the Coconut Grove nightclub burned. This carnage was also due to exits that were inadequate and poorly designed for a place of public assembly. These and other, less disastrous fires have brought changes in building codes and in the public consciousness of fire science, with the result that in the second half of this century so far there have been only two fires with a life loss of more than one hundred persons—the gas explosion in December 1951 in West Frankfort, Illinois, resulting in a life loss of 119 persons, and the Beverly Hills Supper Club fire in Southgate, Kentucky, on May 28, 1977, which took 168 lives.

In the United States in this century, prior to 1950 there were at least 15 fires that took the lives of more than 100 persons. The improvement since 1950 has been because of better fire safety measures. These deal ·with such things as number and location of fire exits, number of persons permitted in a given structure or place at one time, fireproofing of decorations and interiors, installation of automatic sprinkler systems, enclosure of stairway and elevator shafts, and many other aspects of fire safety and prevention. There is no conflict of interest between life safety and protection of property, because every measure that prevents property from burning also contributes to the safety of human beings.

A too-neglected area is the life hazard of fires in dwellings. Loss of life in home fires could be greatly reduced by home fire drills and the widespread installation of reliable home fire alarms.

An NFPA study of 500 building fires in which fatalities were reported indicates that in 30 percent of the fires, the victims could not escape because fire cut off paths to the exits. In another 1.2 percent of the total, there were inadequate exit facilities. The causes of the fires: 56 percent were because of careless smokers setting fire to bedding, upholstery, or clothing. Other sources were heaters (13.8%), electrical wiring or appliances (7.5%), cooking stoves (7.0%), arson (4.2%), and miscellaneous sources (11.5%). The point these figures raise is that most of the deaths could have been prevented with more knowledge of fire science, properly applied.

Property Loss

During the past 25 years alone, United States property loss due to fire increased from $730 million a year to about $4 billion annually. In this day of large figures, $4 billion may not seem much, but consider the following.

Picture a company with capital of $4 billion that started business the year Christ was born. Suppose the business was so poorly managed that the company collected no interest on its reserves and it lost an average of $5000 a day, Sundays and holidays included. That company would still be in business today, and unless its situation got worse, it would continue to operate until the year 2000. At that time its losses over the past 2000 years would equal the value of the property we burn up in this country in one year. In the case of a business losing money, someone else might profit from the company's losses; but in the case of fire, useful property is transformed into smoke and a valueless pile of ashes.

Unfortunately, property loss is not always the whole story. Indirect loss caused by fire may be several times greater than the value of the property burned. For example, when a retail store is destroyed, customers develop the habit of going elsewhere so that if the store is rebuilt, a new clientele must also be built up. In addition, there are no profits during the period of low sales volume and no return on invested capital. If it is an industrial plant, the loss of specially trained personnel and the replacement of complicated machinery may take a long time. Even in a dwelling there are losses that cannot be replaced, such as antiques and objects of art, not to mention items of personal, sentimental value.

And in every case, the city loses the burned property from its tax rolls. The list of indirect losses is long and most of these are not covered by adequate insurance. The most regrettable part is that the great majority of such losses could be prevented.

Analysis of Fire Losses

If we are to do anything about the suffering and waste of life and property caused by fire, we must study the reasons for the losses and then try to discover some solutions. The total number of fire fatalities in the United States increased each year up until 1970; since then, there has been a slight decrease, despite increasing population.

Property losses on the other hand continue to increase (see Table 2). There are many factors to be considered, however, if we are to get a true picture of the situation. Inflation alone distorts the total loss figures; 1950 dollars and 1975 dollars are not equal. If we knew the value of burnable property in the United States we could plot the losses as a percentage of the burnable value and get an accurate picture. Unfortunately, there is no way of estimating burnable value, but the Gross

TABLE 2. Building and Contents Fire Loss Trend in United States 1920 to 1974.

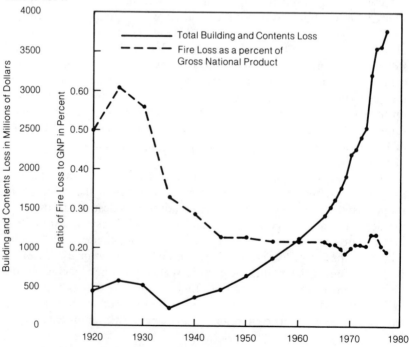

National Product (GNP), or national wealth, has a definite relationship to burnable property. If we plot the annual fire loss as a percentage of GNP we get a reasonably accurate presentation of the property loss situation. A glance at Tables 2 and 3 will show in both graphic and tabular form that the percentage of GNP loss curve is leveling off, but total dollar loss is rapidly rising, and there is room for many more students of fire science to help control these unreasonable losses.

TABLE 3. FIRE LOSS AS PERCENTAGE OF GNP: 1920–1974

YEAR	BUILDING AND CONTENTS FIRE LOSS (million $)	GNP (billion $)	PERCENTAGE FIRE LOSS GNP
1920	447.9	88.9	0.50
1925	559.4	91.3	0.61
1930	502.0	90.4	0.56
1935	235.3	72.2	0.33
1940	285.9	99.7	0.29
1945	484.3	212.0	0.23
1950	648.9	398.0	0.23
1955	885.2	398.0	0.22
1960	1107.8	503.8	0.22
1965	1455.6	676.3	0.22
1966	1528.0	739.6	0.21
1967	1623.0	785.0	0.21
1968	1786.9	864.2	0.20
1969	1933.8	930.3	0.19
1970	2209.2	977.1	0.23
1971	2266.0	1055.5	0.21
1972	2416.3	1155.2	0.21
1973	2537.2	1294.9	0.206
1974	3260.0	1397.4	0.234
1975	3560.0	1528.8	0.233
1976	3558.0	1706.5	0.209
1977	3764.0	1890.1	0.199

Characteristics of Fire

So far, we have briefly traced the history of fire and man's efforts to control it. We have also noted the continuing need for fire science specialists by outlining the toll fire has taken in recent years. Let us now consider the characteristics of fire and heat that are related to fire protection and that we will refer to when we discuss fire prevention, detection, and suppression.

The ancient Greeks believed that air, earth, water, and fire were the four elements that made up the universe and that a god was in charge of each element. Fires occurred when the god of fire became upset over one thing or another. To put the fire out, one prayed to the god of water. If that god was in a good mood, he sent rain and put the fire out—it was as simple as that. Our knowledge of fire is now based on fact and, through research, we are learning more about the principles of combustion and the behavior of fire. The more we learn, the more exact the science of fire prevention and control becomes.

The Classic Triangle Concept of Fire

Fire is defined primarily as rapid oxidation accompanied by heat and light. In general, oxidation is the chemical union of any substance with oxygen. The rusting of iron is oxidation but it is not fire because it is not accompanied by light. Heat is generated, but so little that it can hardly be measured. Burning can occur as a form of chemical union with chlorine and some other gases, but for our present purposes, we need only consider fire that involves oxygen.

Fire can usually take place only when three things are present: *oxygen* in some form, *fuel* (material) to combine with the oxygen, and *heat* sufficient to maintain combustion. Removal of any one of these three factors will result in the extinguishment of fire. The classic "fire triangle" (Figure 2-3), in which each leg represents one of the three factors necessary for combustion, illustrates this. Opening the triangle by removing one factor will extinguish a going fire, and keeping any one factor from joining the other two will prevent a fire from starting.

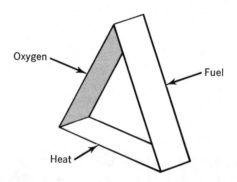

FIGURE 2-3. Fire triangle.

The Tetrahedron Concept of Fire

Recent reasearch suggests that the chemical reaction involved in fire is not as simple as the triangle indicates, and that a fourth factor is present. This fourth factor is a *reaction-chain* where burning continues and even accelerates, once it has begun.

This reaction-chain is caused by the breakdown and recombination of the molecules that make up a combustible material with the oxygen of the atmosphere. A piece of paper, made up of cellulose molecules, is a good example of a combustible material. Those molecules that are close to the heat source begin to vibrate at an enormously increased rate and, almost instantaneously, begin to break apart. In a series of chemical reactions, these fragments continue to break up, producing free carbon and hydrogen that combine with the oxygen in the air. This combination releases additional energy. Some of the released energy breaks up still more cellulose molecules, releasing more free carbon and hydrogen, which, in turn, combine with more oxygen, releasing more energy, and so on. The flames will continue until fuel is exhausted, oxygen is excluded in some way, heat is dissipated, or the flame reaction-chain is disrupted.

Supporting this concept is the discovery of many extinguishing agents that are more effective than those that simply manage to open the triangle. Because of this discovery, we must modify our fire triangle into a three-dimensional pyramid, known as the "tetrahedron of fire" (Figure 2-4). This modification does not eliminate old procedures in dealing with fire but it does provide additional means by which fire may be prevented or extinguished.

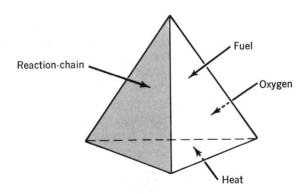

FIGURE 2-4. Fire tetrahedron.

Oxygen

Most fires draw their oxygen from the air, which is a mixture of approximately 21 percent oxygen, 78 percent nitrogen, and small amounts of other gases. Some materials, however, contain enough oxygen in a form that is easily liberated to support combustion without any other source of oxygen. These are known as *oxidizers* and include such compounds as many of the nitrates, chlorates, and peroxides. When present in a fire with burnable material they promote vigorous burning. Finally, pure oxygen, stored as a liquid or as a gas under pressure, can be a source of this fire factor. (Pure oxygen is stored in steel cylinders, such as we see in welding shops, hospitals, and airplanes.)

If a fire burns in a closed room, the oxygen will gradually be used up and the fire will diminish. If no additional supply is available, the fire will go out. However, if a limited but continuous supply is provided (which is often the case), the fire will smolder. Smoldering causes the fuel to vaporize into flammable gases that are only partly burned, since there is not enough oxygen for complete combustion. The room gradually fills with these gases and simultaneously becomes very hot. If air is then allowed to rush into this space, for instance by the opening of a door or window, the oxygen side of our tetrahedron gets a big boost; complete combustion takes place almost instantaneously, causing what we call an explosion. In the fire service, such an explosion is known as a "back draft."

Since oxygen is so readily available from the air, eliminating this side of the tetrahedron to *prevent* fires is usually not practical. In efforts to *suppress* a going fire, however, removing the source of oxygen is a common approach, often referred to as "smothering" the fire. Methods are discussed in Chapter 6.

Heat

Heat for ignition can come from many sources: open flame, the sun, electricity, friction, and so on. The intensity of heat required to start the chemical action of combustion varies with each type of fuel. This intensity, *the ignition temperature, is defined as the minimum temperature to which a substance (fuel) must be heated in order to initiate or cause self-sustained combustion independent of another heat source.*

Most solid materials have an ignition temperature of +204° Celsius to +399° Celsius (+400° to +750° Fahrenheit). These temperatures, however, vary with conditions: time of exposure, size and shape of container, concentration of oxygen, humidity, and others. Wood, for instance, when subjected to 399°C. (750°F.) for a short time, will normally start to burn; but, if exposed to a much lower temperature, say 177°C. to 204°C. (350° to 400°F.) for about half an hour, it will begin to smoke and give off gases that are readily ignited.

For combustion to take place, most substances must be heated rather rapidly. After ignition temperature has been reached, burning will continue as long as the fuel remains above this temperature. The heat to maintain the ignition temperature is usually produced by the chemical reaction between the oxygen in the air and the substance that is burning. The amount of heat produced is called the *heat of combustion.* Heat of combustion also varies with every type of fuel and is usually expressed in Btu (British thermal units). *A Btu is defined as the amount of heat required to raise the temperature of 1.0 pound of water one degree Fahrenheit at sea level.* (In the metric system, heat of combustion is expressed in *joules.*) While this figure is important in determining the amount of *potential* heat in a quantity of fuel, remember that it does not indicate the momentary intensity of the fire as it burns. *Intensity* depends upon the rate at which oxygen is supplied.

Whenever we have a material that needs less heat to reach ignition temperature than the material will produce as heat of combustion, we have the possibility of a self-sustaining combustion.

The heat side of the tetrahedron offers many avenues to prevent and extinguish fires. Preventive methods can be summed up as follows: keep all sources of ignition away from material to be protected, but, when heat is required nearby, apply the proper safeguards to the heating devices. Extinguishment, of course, involves lowering the temperature of the fire to a point below the fuel's ignition temperature. In other words, "cool it." This is most often done by the application of water. We will discuss these methods in detail in later chapters.

Special Cases of Heat and Ignition   A special problem of heat is called "spontaneous heating," a process that increases the temperature of a material without drawing heat from outside sources. If the heating continues until the ignition temperature is reached, "spontaneous ignition" or "spontaneous combustion" occurs.

A few substances that are not combustible themselves may cause heating and, if combustible material is present, start a fire. The most common example of this is unslaked lime. When water is added to unslaked lime, the reaction generates considerable heat: 520 kJ (493 Btu) per pound of lime, to be precise.

Another material that *is* flammable and that is also susceptible to spontaneous heating is *white phosphorus.* This, and a few other related chemicals, will burst into flame when exposed to air, but such materials are found mainly in chemical laboratories. Their characteristics are well known and safeguards are taken.

Sodium and potassium oxidize quickly in air with little heat release. However, contact with water produces an extremely rapid chemical reaction, generating great heat and highly flammable hydrogen, that ignites immediately. Understandably, these materials are rarely handled and only with considerable respect for the hazards.

By far the most common group of materials subject to spontaneous combustion are those that combine with oxygen rapidly enough to allow the heat from the chemical reaction to build up to ignition temperature. Many *organic substances* combine with oxygen from the air and give off heat. Normally, this heat is generated so slowly, or conditions are such, that it dissipates without appreciably raising the temperature. Most of the vegetable oils, such as olive oil, peanut oil and linseed oil, are capable of this kind of heating. They all combine with oxygen and give off considerable heat, especially when a relatively large surface is exposed to the air. Some animal oils such as cod liver oil are also subject to spontaneous heating, but most oils of this type present only a slight danger of heating.

Accidental fires have occurred when a cloth, saturated with a vegetable oil, has been tossed into a bucket or other container. If the cloth were hung on a line, air currents would dissipate the heat; or, if the cloth were rolled into a tight wad, there would not be enough surface exposed to the air to combine with its oxygen. But, when a loose bundle of soaked cloth is protected from air currents, the cloth will often heat to ignition temperatures and burst into flames.

Some materials not subject to spontaneous heating at normal temperatures acquire this characteristic when partially heated from an outside source. Sponge rubber heated in a clothes dryer may continue heating to the ignition temperature. Even wood, when subjected to long exposure to a moderately elevated temperature, such as near a steam pipe, has been known to continue heating and to ignite spontaneously.

A different type of spontaneous heating may occur in stacked hay or other agricultural products, especially when wet or improperly cured. This oxidation usually occurs some weeks after the crop is stacked. Current thinking is that bacterial action causes an initial heating and this is continued by rapid oxidation similar to sponge rubber's secondary self-sustaining reaction.

Fuel

Fuels may be in a gaseous, liquid, or solid state, but combustion normally occurs only when a fuel is in the gaseous or vapor state. Solids and liquids, therefore, must have applied energy, usually heat, to vaporize them *before* oxygen can react with the fuel in combustion. An important exception to this general rule is glowing charcoal that burns without flame. In this case, the solid carbon reacts directly with the oxygen of the air.

Solids   Most ordinary combustible solids are compounds of carbon, hydrogen, nitrogen, and oxygen, along with small portions of other minerals. When free burning in air takes place, oxygen reacts with carbon to form *carbon dioxide* and with hydrogen to form *water vapor*. The minerals and nitrogen compounds usually remain in the solid state as ash. When oxygen supply is limited, as in a closed room with no draft, only partial combustion takes place and toxic *carbon monoxide* is formed instead of carbon dioxide. This toxic gas is the major cause of deaths occurring in fires.

An important factor in the case of ignition is the amount of *surface area* of a given quantity of a solid. Many small pieces of combustible material will heat up much faster than a single large piece. An illustration of this is the way we start a fire in a stove or fireplace: Starting with newspaper (much surface but small particles of solid material) we add some kindling, then larger pieces of wood, finally a log or chunks of coal. All these materials have about the same ignition temperature, but the small pieces of material burn more readily than large objects simply because more surface is exposed per given amount of fuel. This same principle applies when we light steel wool with a match, but we cannot ignite a steel bar with any ordinary heat.

This last example brings up another characteristic of solids that is important in fire behavior. Certain materials, such as most metals, are good conductors of heat, while others such as wood are poor conductors —in other words, good insulators. Since considerable heat is necessary to vaporize the solid so that combustion can take place, the heat must usually be concentrated. If a material is a good conductor, heat is dissipated so fast that no point reaches a temperature sufficient to vaporize the solid.

Moisture content. A characteristic of solid fuel that is related to flammability is moisture content; anyone who has tried to start a campfire with wet or green wood knows this. Moisture content not only affects the ease with which fire may be started, but has considerable influence on the speed of burning. Many of our most common fuels, such as wood and paper, take up water from the atmosphere or other source and hold it, much like blotting paper. Before ignition can take place, this moisture has to be eliminated, usually by vaporization. This requires extra heat, over and above what is needed to vaporize the fuel. But once the fire is under way, the heat increases and speeds up evaporation so that the moisture content becomes less important, although water vapor dilutes the oxygen immediately surrounding the fuel and retards the combustion process.

The importance of moisture content is easily seen when we consider forest fires. Here, long periods of dry weather nearly always result in serious fires. Considerable research on this problem shows that *wood* with a moisture content below 15 percent can be ignited by as small a flame as a match.

Flame spread (the speed with which flame propagates over the surface of a fuel) has special meaning when we consider the materials used for interior finishes in buildings. Fires that spread rapidly over the surface of walls and ceilings have been responsible for many deaths and considerable loss of property in recent years. We measure this characteristic of a material in the following way: The material is installed on the ceiling of a long tunnel called a Steiner Tunnel, and flame is applied to one end. Observers time the spread of flame to a point 30.5 m (100 feet) away. Red oak, fairly constant under this test, is rated at 100; cement asbestos board is rated zero. Other materials are rated lower or higher than red oak, according to the time taken by the flame to spread 30.5 m (100 feet). (See Figure 2-5.)

The smoke generated can be observed in the same test. Smoke is an important product of a solid fuel from the standpoint of human safety.

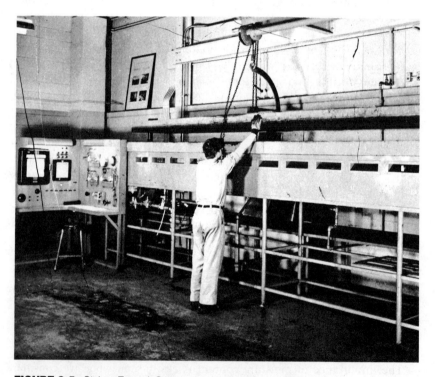

FIGURE 2-5. Steiner Tunnel. Some materials introduce serious fire hazards by their use in building construction. The Steiner Tunnel determines the degrees of hazard by measuring flame spread, how much fuel the material contributes to the fire, and how much smoke is produced. Courtesy of the Underwriters' Laboratories, Inc.

Liquids   By definition, fuels in the *liquid state* flow like water and take the shape of their containers. This is because their molecules move freely among each other, but do not separate from each other unless passing into the gaseous or vapor state. These molecules are always moving and the amount of motion depends upon the temperature of the liquid. Some molecules escape, as gas or vapor, into the space above the liquid; this is *evaporation.*

Motion continues among evaporated molecules. Some of the molecules will strike each other or the containing vessel walls and bounce back into the liquid. In a closed container, if the temperature remains constant, the number of molecules leaving the surface of the liquid will become the same as the number returning, establishing an "equilibrium." The force exerted by the molecules of a liquid in motion is called *vapor pressure.* If the temperature of a liquid is increased, the activity of its molecules also increases, raising the vapor pressure. We can demonstrate this by heating a covered pan of water. As the vapor concentration over the water increases with the temperature of the water, the pressure will soon raise the cover of the pan. Similarly, a boiler can overheat enough to rupture from vapor pressure. This is sometimes called a boiler "explosion," although there is no combustion. If heat is applied to a container of flammable liquid, say, gasoline, the vapors may be ignited by the heat after the container ruptures, but the rupture itself is due to vapor pressure alone.

When a material in a liquid state is heated, it will reach a temperature above which it cannot go and still remain a liquid. This temperature, reached when *vapor pressure equals atmospheric pressure at the surface of the liquid,* is called the *boiling point.*

As we have just seen, molecules in any liquid move about and some break loose from the liquid to become vapor. How fast this takes place below the boiling point is called the *rate of evaporation.* This can be important in knowing how a liquid will behave in a fire, or in storing such liquids to avoid a flammable situation.

Normally, the vapor of a liquid must mix with oxygen from the air to become flammable. But how much or how little vapor is important to fire protection. There is a *minimum concentration* of vapor in air *below which* ignition cannot occur; there is also a *maximum concentration above which* the vapor will not ignite. These concentrations are referred to as the upper and lower limits of the "flammable (or explosive) range," and are usually expressed as percentages by volume of vapor or gas in a particular space. When the percentage of vapor is below the flammable range, it is said to be "too lean to burn"; when the percentage is above the flammable range, it is said to be "too rich."

Testing a liquid's flammability. A standard test procedure, developed many years ago, is used to establish how flammable a given liquid is.

This test consists of heating the liquid slowly and applying a source of ignition to the surface at different temperatures. When the liquid gives off enough vapor to form an ignitable mixture with the air at the surface, a flame will flash briefly across the surface. The lowest temperature at which this occurs is called the *flash point* of the liquid. Do not confuse this with fire point, which is the lowest temperature at which vapors are given off fast enough to support continuous burning. *The fire point is normally higher than the flash point.* When the flash point test is conducted with a closed vessel, it is referred to as *closed cup* (cc) flash point; if conducted in an open container, it is called *open cup* (oc) flash point.

The National Fire Protection Association's "Flammable and Combustible Liquids Code" establishes standards for safe storage and uses of flammable liquids. This code defines a flammable liquid as "any liquid having a flash point below 37.8°C. (100°F.) and having vapor pressure not exceeding 280 kilopascals, kPa (40 psi) absolute at 37.8°C. (100°F.)." Absolute pressure measurement includes 101 kPa (14.7 psi) of atmospheric pressure. The code divides flammable liquids into several classes. We will discuss these later, when we consider the techniques of fire prevention and suppression for these liquids (see Chapters 4 and 6).

Another important property of a flammable liquid is its weight compared to water. This property is called *specific gravity.* The specific gravity of water is conventionally set at one (1.0). Any liquid that floats on water therefore has a specific gravity of less than one; if water floats on a liquid, the specific gravity of the liquid is greater than one. This plays an important role when extinguishing fires fueled by flammable liquids.

The *viscosity* of a liquid is its resistance to flow. Molasses is a highly viscous liquid, while alcohol has a low viscosity. As the temperature of a liquid increases, its viscosity usually decreases; this fact must be kept in mind by fire fighters for reasons that will be apparent later.

The chances of a fire starting in a given flammable liquid and the speed with which it may spread largely depend on the characteristics we have just discussed. Clearly, we need educated care in the management of all flammable liquids, but the requirements are even stiffer when we have a mixture of liquids. A good example is certain cleaning solvents that have a high flash point and appear to be perfectly safe to use. But they are mixtures of gasoline (or some similarly flammable liquid) and carbon tetrachloride, a good fire extinguisher. A fact that the user may not recognize is that carbon tetrachloride has a much higher evaporation rate than gasoline. When the solvent is exposed to the air, the carbon tetrachloride in the mixture evaporates and leaves only the hazardous gasoline.

Gases   We sometimes find it difficult to picture gaseous fuels because many of them are colorless, some are odorless, and none has a definite shape; but they do have certain descriptive features of their own as do solids and liquids. A gas should be thought of as widely separated molecules, speeding straight through space until colliding with each other or the walls of some container. This motion explains why gases tend to fill any space in which they are released. Of course, the larger the space, the less dense the gas becomes.

Gases share many of the characteristics of liquids. The effect of pressure on a gas, however, is notably different than on a liquid or solid, because the spacing and movement of the molecules make it easy to compress a gas. Three basic laws apply to the behavior of all gases:

- Boyle's Law, which says: "When the temperature is kept constant, the volume of a gas is inversely proportional to the pressure upon it." In other words, as pressure is applied to a gas, the volume decreases in proportion. This can be expressed mathematically as:

$$PV = \text{Constant.}$$

- Charles's Law, which describes the effect of temperature, states: "When the external pressure is kept constant, the volume of a gas is directly proportional to its *absolute* temperature." As temperature increases, the activities of the gas molecules become more powerful, increasing the internal pressure on the walls of the gas's container. Expressed mathematically:

$$\frac{V}{T} = \text{Constant,}$$

where T is the temperature.

- The third basic law, having to do with the relation between temperature, pressure, and volume of gases, known as Avogadro's Law, states: "Under equal conditions of temperature and pressure, equal volumes of gases contain the same number of molecules." This knowledge is helpful in determining the relative weight of a gas by a comparison of the molecular weight of the gas to a standard.

For the fire fighter, a related and more useful property of a gas is its *specific gravity:* the ratio between the weight of a given volume of gas and an equal volume of air, free from water vapor and carbon dioxide. Both gas and air are measured at the same temperature and pressure when determining specific gravity, and air has a fixed value of one (1.0). There are tables giving specific gravities for gases. If the figure given for a particular gas is less than one, the gas is lighter than air and will rise. Hydrogen, with a specific gravity of 0.1, is a good example. If the figure is greater than one, we know that the gas is heavier than air and will tend to remain close to the ground. Chlorine is an example of a heavy gas, being two and one-half times as heavy as air.

In recent years, the use of many gases has increased so much that an entire industry has grown up to supply the demand. Because liquid gases cost less to ship than compressed gases (per unit of volume), and because liquid gases take up far less storage space, per unit of volume, gas liquefaction has become big business. Today, most of the gases in common use are found in liquid form.

In simplest terms, we liquefy a gas by subjecting it to extreme pressure, removing the heat caused by compression, and continuing the pressure and cooling until the gas liquefies. Some liquid gases reach temperatures close to absolute zero, $-273.2°C.$ $(-459.7°F.)$; others are not quite so cold but are still many hundreds of degrees below zero.

Main Categories of Gases. Gases are broadly divided into two groups: those that are flammable and those that are not. Notice, however, that while some gases are nonflammable, they support combustion. One such gas is oxygen. It does not burn, but when the supply of oxygen is concentrated, most combustion is intensified.

We use some nonflammable gases that do not support combustion to actually help us extinguish fire. These gases, used to displace the oxygen near the fire, are known as "inert" gases. We will discuss some of them in Chapter 6.

For safety, all gases should be confined in containers or pipes when they are not in use. Containers must be made strong enough to withstand any increase in pressure from an increase in temperature, whether caused by fire or some other form of heat. A common hazard, resulting from our widespread use of gases, is escaping gas that no one realizes is present. The most widespread flammable gases, liquefied petroleum gas or natural gas piped into homes, have a strong distinctive odor added to the pure gas to make sure that a leak will be detected quickly.

Causes of Fire

There is a saying of long standing in the fire prevention field that is something like this: "The three principal causes of fire are men, women, and children." That statement still has considerable significance because most of the more than one and a quarter million building fires that occur every year in the United States are because of human errors either of omission or commission. For that reason considerable importance is placed upon educating the public along fire prevention lines.

For many years we have referred to the "causes" of fire. This was not always strictly correct. For instance, flammable liquids have been listed as one of the causes of fire, yet (as we have seen previously in this chapter) flammable liquids cannot cause a fire unless some ignition source is provided, and one may well ask, "What caused the fire, the flammable liquid or the ignition source?" This is one of the reasons why the last edition of the *Fire Protection Handbook* published by the National Fire Protection Association, instead of listing fires by causes, lists them by "fire ignition sequence."

Table 4 shows in graphic form the annual loss in the United States by the four leading known fire ignition sequences. Perhaps the most alarming thing is the tremendous increase in losses because of incendiary or suspicious fires. As shown on the left axis, incendiary and suspicious fires moved from fourth place in the past to where, in the five-year period 1970–1975, the losses due to these fires far exceeded any other fire ignition sequence. This is likely because of the social unrest in this country during that period and because complete preventive measures are beyond the scope of fire science, but some of the things fire science specialists can do will be discussed in Chapter 4 under "Fire Investigation."

The remaining three sequences shown have developed standard safeguards that range from thick books, such as the *National Electrical Code,* to such simple conventions as keeping matches out of the reach of children.

TABLE 4. Estimated Loss by Fire Ignition Sequence. Courtesy of the Underwriters' Laboratories, Inc.

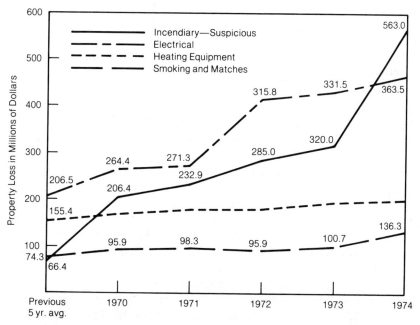

Fire Spread

A small percentage of building fires cause a very large percentage of losses. For instance, in 1976, less than 0.05 percent of such fires caused 18 percent of total property loss. These are termed *large loss fires* and from these comparatively few fires we can learn a lot about why fires spread.

Reasons for the spread of fire are grouped into three categories by the NFPA: structural defects, contents improperly handled or stored, and fire protection defects. Each of these categories is broken down into three or four subdivisions and is presented in tabular form in the *Fire Protection Handbook* published by NFPA. Among other analytical data, the handbook reports the total number of large loss fires in which each of these reasons contributed to the spread of the fire and by which class of occupancy.

We will confine our discussion of the causes of fire spread to only a few of the fundamental contributing factors. Most frequently mentioned structural defects are concealed spaces above or below floors and ceilings, combustible structure or framing, unprotected vertical openings (stairways and elevator shafts), and non-fire-stopped walls.

Most frequently mentioned as contents that influence fire spread are products in storage, and flammable liquids and gases not properly contained.

Related to these hazards are the shortcomings in fire protection measures that are the responsibility of property owners, the most frequent being the lack of an automatic sprinkler system or one that could not control or extinguish the fire.

Conflagrations

The term "conflagration" is often applied to any large-scale fire. More properly, its use should be limited to fires such as those that devastated Chicago in 1874, San Francisco in 1906, and 65 Japanese cities during World War II. Although there is no universally accepted definition, the best practice, according to the NFPA, is to limit our use of the word *conflagration* to fires that extend over a large area, crossing natural or manmade barriers in the process, and destroy many buildings. If we adopt this definition, a fire in a large industrial plant, even though it involved a number of separate buildings and did extensive damage, would not be termed a conflagration unless it spread beyond the limits of the plant. Similarly, a fire that destroyed an entire block of closely packed buildings would not qualify as a conflagration because, again, no barriers were crossed and its extent was limited.

When a true conflagration exists, it presents special problems of prevention and suppression. Analysis of past fires that reached conflagration proportions reveals the presence of one or more of the following combinations of circumstances:

1. There was a wind blowing at 48 km per hour (30 mph) or more.

2. The fire began in a place where the danger of fire is always high, commonly called a fire-breeder, located in a congested section that lacked protection for external exposures. It spread rapidly in one or more directions before effective resistance could be marshaled. Once started, it moved with the wind and crossed streets by radiated heat.

3. The fire began in one of many closely built residences and spread rapidly to others because of combustible construction and wooden shingle roofs. Inadequate fire protection forces and inadequate water supplies made control impossible.

4. An extensive forest or brush fire swept down on a town or city, attacking its buildings on a wide front.

5. One or more explosions, hurling burning materials, hot metal, or similar sources of ignition, or generating waves of hot gases, started a number of fires simultaneously over a wide area.

Under any of these circumstances, the major factors in a fire reaching conflagration proportions are:

- The enormous increase in the rate of fire propagation that a high wind causes.

- The spread of fire by sparks, embers, and burning brands, borne aloft for hundreds of feet and dropped on combustible roofs or other combustible materials.

- Travel of superheated gases of combustion before bursting into flame.

- Ignition of exposed buildings by radiated waves of heat.

Once a fire reaches conflagration proportions, it can sustain itself as long as there is combustible material within reach. This is easily understood: strong convection currents fairly suck in oxygen-laden air, there is more than enough heat present, and, given fuel, the classic fire triangle is complete.

Like most fires, conflagrations are difficult to extinguish once started. But, by studying past conflagrations and applying fire protection principles, we may be able to prevent future disasters. In fact, such studies became the basis of what we know today as the Grading Schedule for

Municipal Fire Protection (published by the Insurance Services Office, 2 World Trade Center, New York).

As might be expected in a free enterprise system such as ours, it was private enterprise, the insurance business, that made the first major effort to develop fire protection and prevention methods. Formed in July 1866 by 75 fire insurance companies following a conflagration in Portland, Maine, that put many firms out of business and severely shook public confidence in fire insurance, the National Board of Fire Underwriters first began to put the industry's house in order. By 1874, the Board and its member companies were giving weight to fire prevention measures in fixing rates with the same consideration extended to fire protection shortly afterwards. (See Chapter 3 for detailed discussion.)

Studies of conditions favoring or preventing great fires got a real boost right after the Baltimore disaster in 1904 when some early fire scientist pointed out that nothing had happened during the fire that a study of conditions could not have foretold. Members of the Board decided to make such studies in the larger cities of the country and hired a crew of engineers to make reports. A 1905 report, available to members of the Board six months before the San Francisco conflagration, did a great deal to establish faith in this new approach to the problem of fire in cities. The critical question in fixing insurance rates and writing fire insurance policies became: What kind of fire defenses exist in any given city, town, or area?

Inspections, made by teams of engineers working out of headquarters in New York City or branch offices in Chicago and San Francisco, formed the basis of a set of standards for municipal fire defenses. These standards were first published in 1916 as a schedule by which cities could be graded and their relative positions established according to the efficiency of their fire defense systems. Cities are divided into ten classes according to a detailed grading system. This schedule is considered so important to fire science that we will devote all of Chapter 10 to a close study of what it covers.

REVIEW QUESTIONS

1. Describe some of the first methods of starting a fire.
2. What were the "Vigiles" and how were they used?
3. What was the first fire organization in America; when and where was it formed?
4. What is a fire mark and how was it used?
5. Describe a bucket brigade.

6. When was the pump first developed?

7. When were public water systems first used and how were they used for fire protection?

8. Describe the advancement of the pumper up to modern times.

9. How did the chemical engine work and what is used in its place today?

10. Describe the changes in fire department organizations in this century.

11. Where does fire rate among the causes of accidental deaths in the United States?

12. In what type of occupancy does the greatest number of deaths due to fire occur?

13. Name two ways loss of life due to fire in the home can be reduced.

14. What is the approximate annual property loss due to fire in the United States?

15. Is it increasing? Decreasing? Staying the same?

16. What is the effect when losses are plotted as a percentage of the Gross National Product?

17. What is the fire triangle and how can it be used in fire extinguishment and fire prevention?

18. Explain the fire tetrahedron concept.

19. Air normally contains what percentage of oxygen?

20. What is ignition temperature? Heat of combustion?

21. What is spontaneous combustion?

22. Name three characteristics of solid fuels that are important to fire behavior.

23. Considering fuels in the liquid state, what is vapor pressure? Boiling point?

24. What is meant by the phrase "too lean to burn"? "Too rich to burn"?

25. What is specific gravity? Flash point? Viscosity?

26. Considering gaseous fuels, what is Boyle's Law? Charles's Law? Avogadro's Law?

27. What is the specific gravity of a gas? What is an inert gas?

28. Name the three principal causes (fire ignition sequences) of fire in the United States.

29. Name the three principal reasons for fire spread causing large loss fires.

30. What is a conflagration?

3 Fire Protection Services

In Chapter 2, we discussed some of the main characteristics of fire from the standpoint of fire protection. Let us leave the fire burning, so to speak, and take a look at some of the agencies involved in the science of fire protection. Those who are unfamiliar with this field may think that fire science is useful only to fire fighting organizations, but there are many other organizations concerned with fire and fire prevention. Specialists in fire science are not limited to those serving in fire fighting forces. Some of the other organizations that develop our knowledge and apply it to reduce loss of life and property are the subjects of this chapter.

Public Organizations:
Federal Government Agencies

Department of Commerce (DOC)

Many departments of the federal government are involved in fire protection in various ways. The National Bureau of Standards (NBS), Department of Commerce, probably best known for developing a wide variety of standards of measurement used in almost every industry, contributes greatly to fire science. Research by the Building Technical Division of the NBS has made outstanding contributions, such as determining the fire resistance of specific walls, partitions, columns, and floors. The NBS has also obtained valuable information about the effectiveness of some types of fire extinguishers.

The federal government began taking a markedly increased interest in the national fire problem in 1967, and the Congress of the United States passed the Fire Research and Safety Act of 1968, which created the National Commission on Fire Prevention and Control. At that time Congress declared that the growing problem of loss of fire and property from fire was a matter of grave national concern and that there was a clear and present need to explore and develop more effective fire control and fire prevention measures throughout the country. The National Commission on Fire Prevention and Control was appointed by the president and consisted of 20 people versed in various aspects of fire science.

The Commission was required to undertake a comprehensive study to determine practicable and effective measures for reducing the destructive effects of fire throughout the country. After nearly two years of work, it submitted its report to the president on May 4, 1973. The report is titled "America Burning," and includes a comprehensive analysis of the fire problem along with 90 recommendations. As a result of that report, the Federal Fire Prevention and Control Act of 1974 was passed by the Congress and was signed by President Ford on October 29, 1974. The act created the National Fire Prevention and Control Administration (NFPCA) in the Department of Commerce. The act concurred with the basic findings of the National Commission and required the NFPCA to execute a national program aimed at reducing the nation's fire losses. The act states that the NFPCA is to execute "a coordinated program to support and reinforce the fire prevention and control activities of state and local government" and to "supplement existing programs of research, training, and education, and to encourage new and improved programs and activities by state and local governments." It is clear that it was not the intent of Congress to have the federal government take over or direct fire prevention and control activities, but to encourage and support local fire protection programs.

In 1978, Congress extended the life of the NFPCA with additional authority to investigate fires in state and local jurisdictions. It also changed the name of the agency to United States Fire Administration (USFA). Upon recommendation of the President, the agency was transferred from the Department of Commerce to a new department, the Emergency Management Agency. This new agency will deal with various aspects of emergency management besides fire, for example, civil defense and disaster relief. It is understood that this change will not result in the diffusion of USFA into other components of the new agency and will not lessen either the funding for or the impetus of its program.

Although the USFA is a relatively new agency it has made some important starts to accomplish the objectives set forth by Congress. First, it was important to obtain information regarding fires so that a complete analysis of physical and social factors that contribute to life and property loss could be considered. To accomplish this the National Fire Data Center has been created within the USFA. Data are obtained from several sources and incidents reported on standard report forms based upon the Uniform Coding Standard (NFPA 901). It is expected that this will help to measure the problem and lead to some solutions.

The United States Fire Administration (USFA) is also working with the Fire Research Center of the Bureau of Standards and other research agencies to accelerate efforts to transmit fire research products and new fire technology to potential users. Progress has been made in developing a comprehensive master planning procedure for communities that defines the fire risk and develops a method of determining what risk is acceptable. A community fire protection master planning manual has been developed, which should aid local officials to develop their own effective plans of action.

The National Academy for Fire Prevention and Control, an arm of the USFA, is conducting a comprehensive assessment to identify the fire safety disciplines necessary to carry out various tasks. At the present time four principal fire service disciplines have been identified: (1) fire/arson investigator, (2) public fire safety education specialist, (3) fire inspector/code enforcement specialist, and (4) fire fighter. In cooperation with the Professional Qualifications Board of the Joint Council of National Fire Services Organizations, the National Fire Academy is conducting a thorough analysis of the knowledge and skills pertinent to an effective fire prevention and control effort. The plans of the National Fire Academy include the establishment of an educational and training center similar to the National Police Academy operated by the F.B.I. A site for this purpose is under study, but implementation of the plan has been delayed because of lack of funds. It is expected that funds will be available and the academy will become a reality soon.

To give special emphasis to public fire education programs, the USFA has created the Public Education Office. The mission of this office is to assist the fire service and public at large in the planning and implementation of public education programs. The education office with the National Fire Academy is developing a training program for public education specialists.

No doubt many other programs will be developed by the United States Fire Administration as it becomes better established, but it is interesting to note how much has been accomplished in the short time since the federal government became prominently involved in the fire problem.

Department of Transportation (DOT)

The Department of Transportation was organized under the Transportation Act of 1966 and was later given authority over several fire safety programs under the Transportation Safety Act of 1974 (and several other acts passed previously), which authorized the DOT to take over the regulation of fire safety programs in all types of transportation.

The Federal Aviation Administration (FAA) establishes fire and safety standards for aircraft and airports and investigates all fires and crashes to determine causes and loss.

The National Highway Traffic Safety Administration (NHTSA) establishes and enforces safety regulations for automobiles, trucks, and buses. It also administers the DOT Emergency Medical Services Program to serve highway accident victims.

The U.S. Coast Guard (USCG) is very important in establishing standards for merchant and pleasure craft as well as offshore oil drilling rigs. The Coast Guard also provides assistance in fighting all types of marine fires in port facilities.

The Materials Transportation Bureau (MTB) has taken over much of the responsibility formerly assigned to the Interstate Commerce Commission (ICC). It has regulatory powers for the transportation, storage, and labeling of hazardous material and also regulates hazardous materials in transportation pipelines.

Department of the Interior

The Department of the Interior, through the National Park Service, maintains our national parks and protects these valuable areas from fire. Fire regulations, including a complete code for buildings in the parks, are enforced by park superintendents and rangers. The Park Service is responsible for fire fighting operations in the wild lands, but concession operators of hotels, stores, and amusements must maintain structural fire protection. Larger parks require organized fire departments to serve built-up areas.

An important contributor to fire science in the Department of the Interior is the Bureau of Mines. The Bureau maintains an experimental station in Pittsburgh, Pennsylvania, and has developed valuable information on such hazards as dust and gas explosions; explosive limits; static electricity; and explosive and combustible metals. In addition, the nation's fire fighting forces have profited by the Bureau of Mines' work on respiratory equipment and other protective devices.

Department of Defense

All of our armed forces maintain fire fighting and fire prevention organizations. The Navy became particularly interested in fire protection early in World War II when it realized more ships were being lost at sea because of fire than from enemy action. The Navy commissioned trained fire officers and set up schools to train all personnel in fire fighting. Fire protection engineers were assigned to all naval districts to supervise construction and to advise on all fire protection problems. This policy is still in force, and fire protection engineers continue to work closely with base fire chiefs. The Navy also maintains three laboratories that conduct extensive research on fire protection problems.

The Army's fire fighting and fire prevention forces are usually made up of civilians under the general supervision of a commissioned officer, designated as fire marshal. An Army research and development laboratory at Fort Belvoir, Virginia, has contributed notably to developing fire fighting techniques and equipment.

General Services Administration

The General Services Administration, Fire Prevention Branch, is responsible for fire safeguards for U.S. Government buildings in Washington and elsewhere. Sound regulations for construction and fire prevention are closely followed. The Federal Fire Council sees to it that these regulations are followed by a system of fire inspections and detailed fire reports. The council also serves as a forum for interdepartmental discussion of fire protection problems applicable to government buildings.

Department of Housing and Urban Development (HUD)

The Housing and Community Development Act of 1974 provides the authority for HUD to establish and enforce minimum property standards covering one- and two-family dwellings as well as multifamily dwellings. A subsequent amendment to the Act provides for establishment and enforcement of standards covering mobile homes. These standards closely follow those of NFPA, with some amendments.

HUD is an important enforcement aid since its insurance on mortgages and loans and direct-assistance programs is contingent upon compliance with these standards.

Another arm of HUD important to the fire protection problem is the Federal Disaster Assistance Administration, which provides assistance to public agencies and individuals that have suffered damage because of major disasters declared by the president, including fires and explosions.

Consumer Product Safety Commission (CPSC)

This is a relative newcomer to the federal agencies involved in the fire problem, but a very important one. The Consumer Product Safety Commission (CPSC) administers the Federal Hazardous Substance Act, the Flammable Fabrics Act, and the Poison Prevention Packaging Act, with authority to ban hazardous products from the marketplace. It also carries out research programs to support its standards.

Occupational Safety and Health Administration (OSHA)

When the Occupational Safety and Health Administration (OSHA) was established in 1970 within the Department of Labor (DOL), it was highly controversial because of its broad authority. It was authorized to promulgate and enforce mandatory standards providing for the safety and health of employees in their place of work. The law gave OSHA the authority to close down a plant or place of business in the event of noncompliance. As time went on and experience was gained, the emphasis of OSHA changed from punitive measures to consultation and prevention. Some 50 NFPA standards were adopted by OSHA by reference, giving them the force of law. Many problems presented themselves, partly because of the wording of the standards, which is not always acceptable as a law or ordinance. These are being worked out through the cooperative efforts of OSHA and NFPA.

Enforcement of the standards is being financially supported by OSHA and carried out through state officials. A training institute is maintained near Chicago for federal and state enforcement officers and industry personnel. Field training programs and short courses in colleges and universities are other aspects of OSHA's educational program.

Department of Agriculture

Probably the best known, and certainly the largest, fire protection force in the federal government is the National Forest Service, a part of the U.S. Department of Agriculture, which was established and made responsible for the protection of our national forest lands in 1905, four years after the creation of the Yellowstone Park Timberland Reserve, our first national forest (now a national park).

Over the years, more acreage was added to the national forests. Most was government-owned timberland, but some acreage was added through purchase and gifts. Today, there are 154 national forests, 19 national grasslands, several experimental forests, and other projects, totaling more than 200 million acres. This sizable portion of the United States is under the protection of one organization, the Forest Service.

If fire protection were the Forest Service's only responsibility, this would be a formidable enough task, but the Service has many other responsibilities. Two divisions of special interest to us are fire management and forest protection research. (See Figure 3-1.)

Forest Service Fire Protection Operations   The most recent available Forest Service records show a 1975 total of 10,855 fires that destroyed more than 200,000 acres of land under Forest Service protection. Nearly 6000 of these fires were caused by man. These figures illustrate why the Forest Service is so concerned with fire control.

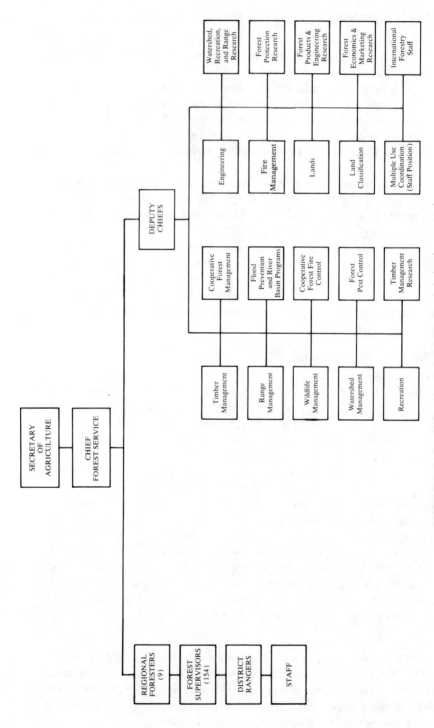

FIGURE 3-1. Forest Service responsibilities and chain of command.

Operations and equipment used in forest fire fighting differ in many ways from metropolitan fire department practices because of the great distances to be covered, inaccessibility, and lack of readily usable water. In spite of these differences, the fundamental principles of fire prevention and extinguishment still apply. Since the occurrence of forest fires is highly seasonal, most fire fighters are hired for the "fire season" only, but all are under the direction of experienced fire officers.

Regional Fire Management Officers usually act in three principal ways: as administrators, directors of field coordination, and supervisors of air operations and training. This last activity is also found at the district level. Aircraft are used extensively to direct fire fighting operations, to carry fire fighters to the scene of the fire, and to bombard the fire with suppressants. Some of the ground equipment will be discussed in Chapter 5.

The National Forest Service's nine experimental stations, and its forest products laboratory in Madison, Wisconsin (under the general supervision of the Forest Protection Research Division), have made important contributions to fire science; notably, developing ground cover less flammable than most ordinary shrubbery, and developing chemical treatments that make fuels fire retardant.

Other National Organizations

The following quasi-governmental organizations operate under charters granted by Congress and operate primarily in the research area of fire problems: the National Academy of Sciences, the National Academy of Engineering, and the National Research Council. These organizations have various committees and they not only report on research but stimulate studies in other areas.

The foregoing is not a complete list of United States Government organizations involved in the fire protection problem, but we have tried to mention the agencies most often referred to. For more information, see *Fire Journals,* Volume 69, No. 6, pp. 51–54, and Volume 70, No. 1, pp. 29–31. This information was later included in the *Fire Protection Handbook* 14th edition, published and copyrighted in 1976 by the National Fire Protection Association. The authors have used, with permission, these and other sources in preparing the above outlines.

Two other organizations, although not part of the federal government, should be mentioned here because they are national in scope and are composed of public employees.

The International Association of Fire Chiefs maintains headquarters in New York and holds annual meetings to discuss fire chiefs' problems. The printed records of these meetings, as well as the association's newsletter, contain much valuable fire science information. The association also sponsors an annual training course in fire administration that is growing in value.

The International Association of Fire Fighters (affiliated with the A.F.L.-C.I.O) is the national representative of local fire fighters' unions. In addition to its concern for the personal welfare of its members, the association is active in promoting fire prevention and training programs. It publishes a monthly magazine that helps keep fire fighters up to date.

Principal Canadian Agencies

In Canada, activities similar to those of the United States National Forest Service are assigned to a Director of Forestry in the Canadian Department of the Interior.

There is also a Dominion Fire Commissioner who is responsible for the enforcement of the Criminal Code of Canada regarding arson and negligence resulting in fires. The Commissioner, an official of the Department of Public Works, directs the fire safety program for the property under the department's jurisdiction. The Commissioner also carries out a fire prevention educational campaign and cooperates with the efforts of Provincial Fire Marshals.

State and County Fire Services in the United States

State and county fire services vary from area to area because of economic conditions, and frequently because of differences in opinion as to how fire protection should be organized. We will discuss the most common and important services in order to understand how they are organized and what responsibilities they have.

Office of State Fire Marshal

According to a recent survey by the Fire Marshals' Association of North America, 46 of the 50 states have the office of fire marshal. In the four states that do not have this office and in some that do, statewide fire protection responsibilities are divided among several agencies, such as Public Safety, State Police, and Attorney General.

The first state fire marshal's office was organized at the beginning of this century. Its chief purpose was to maintain fire records and to investigate fires where arson was suspected. Probably for these reasons, a large number of marshals' offices were made a part of state insurance departments. This is the case in 16 states, and indeed, in many of these states the Insurance Commissioners are fire marshals, although they ususally have deputies specifically trained in fire protection.

In seven states, the fire marshal's office is separate from other departments of government, and the marshals report directly to the governor. When organization takes this form, there are usually advisory boards or commissions, composed of fire protection leaders appointed by the governor, to provide guidance in statewide fire protection.

The responsibilities and organization of the office of fire marshal vary from state to state. In general, most offices investigate and report on fires, particularly those of a suspicious nature. To prevent hazards, the fire marshal inspects all state properties and places of public assembly or institutional occupancy, such as schools, hospitals, prisons, and mental institutions. In about half the states, marshals inspect and report on industrial, mercantile, and other properties, especially where they are beyond incorporated city limits. Some marshals pass on plans for new buildings for public occupancy before building permits are granted.

Increasingly, the state fire marshal has been given responsibility for the control of hazardous conditions that cannot be regulated effectively at a local level. A good example is the transportation of flammable liquids, explosives, and liquefied petroleum gases. Regulation by each city or county is not practical since trucks move from place to place; statewide regulation seems the best answer to the problem.

More than forty state fire marshal's offices enforce regulations governing automatic alarms and sprinklers. This requires people who are specially trained in the engineering of these systems. In recent years, nine states have assigned the job of regulating the servicing of fire extinguishers to their fire marshals.

When a state legislature passes a law establishing the state fire marshal's authority, the law seldom goes into detail. Usually it is an "enabling act," authorizing the fire marshal to draw up suitable rules and regulations that then have the force of law. This is called "delegation of authority" and is done because the marshal and his staff have the special knowledge required to draw up these rules and regulations. The legislature does not often go into the details of regulation, even though there are such national standards as the National Fire Protection Association's available for reference and guidance.

Twenty-eight state fire marshals direct public education programs on fire prevention. (Fire service training, at the state level, is usually conducted by the Department of Education.) Some other responsibilities that most marshals have are regulation of fireworks and advisory services for local fire departments.

The size of the technical staff in state marshals' offices varies from two to sixty-five. Considering the office's variety of duties, it is not surprising that often state fire marshals feel that they do not have enough people to do a satisfactory job. Some offices partly relieve this shortage by deputizing local fire chiefs and fire marshals. This arrangement gives local officers authority to enforce state laws and, when proper communication is maintained, avoids duplication of effort.

Canada's Fire Marshals

In Canada, each province or territory has either a fire marshal or a fire commissioner. The two territories, Yukon and Northwest, share one fire marshal, who is a member of the Dominion Fire Commissioner's staff.

Among the provinces, there is no uniformity in organization. In four provinces, the office is under the Attorney General or Department of Justice; in three, it is a division of the Department of Labor; and in the remaining two, one is under the Provincial Secretary, the other is included in the Department of Public Works. In some provinces the job is called "provincial fire commissioner" and in others, "provincial fire marshal."

The provincial statutes that outline the responsibilities of fire marshal or commissioner are more uniform than the governing state statutes in the United States, but their operations differ to some extent. In all provinces, the office has five important functions:

- Fostering fire prevention.
- Inspecting all properties and directing remedies for conditions likely to cause fire. There is one important exception: responsibility for inspection of *federal property* is given to *federal officials.*
- Investigating or inquiring into fires, when necessary, to ascertain the source, origin, and circumstances.
- Collecting and disseminating information on fires.
- Providing training facilities and advice to establish and administer fire brigades and departments.

Activities from office to office differ as they do in the United States. For example, the fire marshal enforces regulations on flammable liquids in six provinces, liquefied petroleum gas in four, and fire extinguisher servicing in two. Seven offices review construction plans of province-owned buildings, institutional buildings, and places of public assembly.

Technical staffs of provincial fire marshals vary in size from one to eighty-five. All provinces, however, provide for delegating someone to assist at the municipal level. Usually, it is the fire chief. From province to province, the job of the fire marshal is more uniform in organization and activities than is the job of fire marshal from state to state in the

United States. In general, the office provides a focal point for all the fire service in each Canadian province.

State Forestry Departments

Approximately half of the 50 states in our nation have organizations responsible for suppressing fire in areas outside the national forests or the jurisdictions of local fire protection agencies. These organizations are the *state forestry departments.*

These departments have different names in various states, but are always responsible for protecting forest lands within the states, particularly brush and grasslands that are important watersheds. In some states, the forestry department's responsibility extends to all lands not otherwise protected, although the county usually bears the cost of fire suppression for any such lands.

Forestry departments vary greatly in size and organization. Largest is the California Division of Forestry (C.D.F.). A brief look at that organization will give us a good idea of how all of them work. (See Figure 3-1.)

The C.D.F., a subdivision of the California Department of Conservation, maintains an elaborate radio network for communication and dispatching. The staff includes a fire prevention section and some arson investigators. In case of large fires, the State Forester has the authority to impress local citizens as fire fighters. However, this has been a poor source and has fallen into disuse. Instead, extra workers are drawn from state prisons or local jails. The size of forest service fire crews and types of equipment depend on the amount of fire protection required within a given county. Any county may expand its protection beyond that which the state normally provides by cooperative agreement with the C.D.F.; the county agrees to pay the state the cost of maintaining additional protection. In some counties, the county fire organization takes over complete fire protection responsibility and the state reimburses the county. The State Forester may also make agreements with the National Forest Service to shift the responsibility for fire protection in certain areas.

State Disaster Defense

During World War II, every state had an office of civilian defense. Many have kept the office at the state level, now usually called the Office of Emergency Government or Emergency Services, as the key control point in plans for handling major disasters. Currently, the authority of such an office depends upon a "disaster act" (or some similar law) that gives a state's governor (or a disaster council) extended power in an extreme emergency. The coordination of emergency services and the preparation of plans for unified action are the principal tasks of the office.

The director must view the state's problems in much the same way as the governor does, for the director may have to act in place of the governor during an emergency. Assistants manage each of the office's separate service areas, such as fire, law enforcement, rescue, and medical. (See Figure 3-2.) To make a disaster plan work, local agencies throughout the state must cooperate with the state office.

Some states have set up radio communication networks to allow stricken communities to call for help when telephone lines are down. The same system is available for dispatching workers and equipment to areas in need. At least one state keeps a large number of special fire trucks on continuous standby at local fire departments so that they can be moved into a disaster area on short notice.

Local fire control officials must be thoroughly familiar with the state fire disaster plan since any community may suffer fires beyond the control of its own fire forces. Disaster resources must not, however, be depended upon to take care of fires that can be expected under normal conditions.

State Fire Training and Education Programs

Nearly all states have some form of fire training program. In some states, it is conducted by a state fire fighters' organization; in others, it is handled by the state college, state university, state department of education, or a combination of all three. Such training is important, of course. Even with the best equipment, someone just "taken off the street" cannot provide us with adequate fire protection; *it takes thorough training and education to develop effective fire fighters.*

In some states, paid fire departments are given a week or so of intensive course work, ususally at a state university, focused on fire training.

Some states have used a very effective system for training fire fighters called the "itinerant instructor program." This consists of one or many well-trained fire fighters with a background in fire science traveling throughout the state doing training in departments that have filed a request for this service. The instructors use the apparatus and equipment in the department and demonstrate its most effective and efficient use. This program is especially good for volunteers or small departments because classes can be held at night for volunteers and without taking people or equipment out of service.

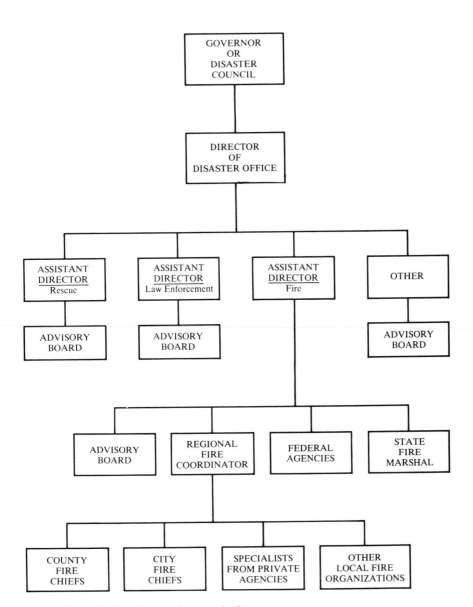

FIGURE 3-2. Typical disaster-plan organization.

A more sophisticated state training program uses a fire training academy, which provides dormitories and boarding facilities as well as an assortment of training aids such as towers, smoke rooms, and various types of apparatus and equipment. The cost of providing such a facility and providing an adequate staff and maintenance can run into high figures. An additional cost is borne by the cities that send firemen to the academy in the form of board and room, at the same time paying salaries for these fire fighters who are not on duty. A program of this kind needs a broad base such as a National Fire Academy, where persons can be trained to come back as instructors for the rest of the department.

We need not consider the fire training program in every state, but it seems quite clear that regardless of geographical or political location, the greater the opportunity for more fire education, the greater the demand.

County Fire Protection Agencies

As a rule, fire protection agencies do not have a big part in county government. But we must not overlook the important exceptions. Some counties do maintain well-equipped and well-staffed fire departments in order to provide fire protection, under contract, for incorporated cities within the county. These organizations are similar to the municipal fire defense systems that we will consider in detail later.

Some counties that do not maintain a countywide fire fighting force may still have a fire prevention bureau to enforce county fire ordinances. These bureaus, usually headed by a county fire marshal, operate very much like the fire prevention bureau of a city fire department.

Most states have laws that cover the organization of fire protection districts. These districts are usually formed at the county level and may include property not otherwise protected. Such districts are usually administered by a board of fire commissioners, who are either appointed by the county's governing body or elected by the people within the district. The responsibilities of these organizations may include the protection of brush and grasslands, grain fields, rural buildings, or entire municipalities within the district.

Private Fire Protection Agencies

The development of fire science is by no means limited to government. Private fire protection agencies play an important part in the field of fire protection. They develop standards and rules of safe practice, conduct research, test materials and appliances, collect statistics, and circulate information on fire safety matters. We will limit ourselves to a brief look at the history, organization, and accomplishments of some of the better known agencies, remembering that there are many others.

The National Fire Protection Association

The best known fire protection organization is the National Fire Protection Association (NFPA). Organized in 1896 by 18 people who were mostly from the fire insurance industry, the NFPA had three aims: (1) promotion of the science and improvement of fire protection methods; (2) preparation and circulation of information; and (3) cooperation of members and the public in establishing proper safeguards against loss of life and property by fire. The Articles of Organization of the NFPA still carry these three aims as objectives.

The problems that the small group faced in the late nineteenth century dictated the way their organization grew. Then, just as now, carelessness and ignorance of fire hazards were the principal causes of fires. Education was the only answer to this problem. Electricity, just coming into general use, created a major new hazard and rules for safe installation lagged behind its use. Although there were automatic sprinkler systems, reliability was poor. Standards for safe use were needed to meet these and similar problems. But in order to formulate proper standards, the NFPA needed detailed facts about important fires.

The need for fire information, safety standards, and public education was the main force in shaping the NFPA as it is today. Recognized as the international clearinghouse for fire safety information, the NFPA is a nonprofit corporation, headquartered in Boston, with over 32,000 individual memberships which total increases at an average annual rate of 8 percent and more than 200 organization members. Every state in the United States and every province of Canada is represented in the membership, as are some 79 other countries.

While the original 18 members were mostly in the insurance business, that industry's representation has dropped in recent years. The largest portion, 40 percent, has been from commerce and industry (including 11% in insurance); the fire service has accounted for 32 percent; hospital administrators make up another 11 percent; and the remaining 17 percent come from local, state, and federal governments, related professions, such as architecture and engineering, and many other backgrounds.

The organization is in the charge of a board of directors, consisting of elected officers, including two past chairmen of the board, and 18 members elected by the Association for three-year staggered terms. The day-to-day affairs of the NFPA are administered by the president and a paid staff of about 200 working in the Boston Headquarters. Several members of this staff enjoy worldwide reputations in their particular fields of fire science.

The most important NFPA function is developing fire safety standards. These standards are *advisory rules* covering the many different areas and activities that have an element of fire hazard. Although the NFPA's rules are widely adopted and incorporated into the laws of the states and the ordinances of counties and municipalities, the Association itself considers its standards as simply advisory.

NFPA standards are the result of the conclusions of 165 technical committees, made up of 2400 experts in special fields who serve the NFPA without compensation. Committee members are drawn from businesses and other interests affected by NFPA standards. When a committee agrees on a new proposed standard, or a change in an old one, the proposal is printed and distributed to all members before it is submitted to the annual or fall NFPA meeting for hearing. At the annual meeting, the full NFPA membership present may adopt the standard, revise it, or return it to the committee for change or discard.

The actual way in which standards are prepared has developed over the years. Obviously, its success depends greatly on the people who serve on the committees. The board of directors selects technical committee members after careful screening by the NFPA staff and the Standards Council. Nearly 250 standards have been adopted and published in 16 volumes. Some of the better known are the National Electrical Code, the Safety to Life Code (formerly Building Exit Code), the Flammable Liquids Code, and Standards for the Installation of Sprinkler Systems.

In general, the NFPA's method has worked well. It has produced standards that have helped to assure reasonable fire safety without prohibitive expense, interference with established industrial processes, or undue inconvenience. The best indication of their practical value has been their wide adoption by legislative bodies, insurance organizations, and trade associations.

Besides the publication of its standards and Fire Codes, the NFPA publishes many technical books, pamphlets, and periodicals. The *Fire Protection Handbook* is widely recognized as the authoritative reference on fire protection, and more than one hundred other books on fire protection subjects are published by the NFPA. The periodicals range from "Fire Technology," designed to cover technical subjects in the field of fire protection engineering, to the "Sparky Fire News" and comic books, written to attract the interest of schoolchildren. The "Fire Journal," published bimonthly, is distributed to all members of the Association, but it is not for sale. It contains articles of general interest on fire matters as well as minutes of the board of directors' and Standard Council meetings.

The NFPA is involved in applied research programs, many of which are funded by government agencies. These include development and implementation of training programs and fire incidents and prevention inspection reporting programs.

A primary function of the NFPA since its organization in 1896 has been the recording and analysis of fire experience. The Association's Fire Information and Systems Division has accumulated something like a quarter of a million fire reports and adds to this at the rate of about 10,000 reports each year. Records are kept on all fires with more than $100,000 loss. Special reports are published on those of over $250,000 loss or involving a loss of life. A careful analysis of these fires indicates where standards may be weak, where public protection broke down, or where another hazard may exist. Frequently, these studies show an exceptionally large number of fires in a certain type of occupancy, which justifies the publication of an "Occupancy Fire Record." Some 50 of these reports are available, as well as special reports on fires of particular interest, fire loss statistics, and other information pertaining to fire records.

Administration   Although the affairs of the Association are under the charge of the board of directors, most of the activities are the responsibility of the staff. The chief executive officer of the staff is the president, who is elected by the board of directors and serves at their pleasure. The president is an ex-officio member of the board. The vice president is also elected by the board upon recommendation of the president. The organization of the staff has been revised many times to keep up with fire protection needs of NFPA members and the public.

The functions of the staff are divided among eight divisions as follows. The Business Services Division handles advertising and membership and subscription promotion. The Business Systems Division takes care of publication and general promotion. The Editorial Division is responsible for the development and production of publications, assisted by the Graphic Arts Department. The Engineering Division handles all the fire protection specialties including the field services and an Educational Technology Unit. The Fire Information and Systems Division handles fire analysis and records. The Public Affairs Division takes care of public and press relations. The Public Protection Division is responsible for matters dealing with fire departments and water systems and provides liaison with technical committees on these subjects. The Research Division organizes and implements the applied research programs. All divisions work closely with each other and with the technical committees.

Because of diversity of interest and the time limits annual meetings impose on covering all subjects of interest, the Association has set up seven sections, each of which holds an independent meeting to transact the business of the section and discuss problems of particular interest. These seven sections consist of the Chief Electrical Inspectors Section, the Fire Science and Technology Educators Section, the Health Care Section, the Fire Marshals' Association of North America, the Fire Service Section, the Industrial Fire Protection Section, and the Railroad Section.

Society of Fire Protection Engineers (SFPE)

Organized in 1950, the Society of Fire Protection Engineers is the professional society of engineers involved in the multifaceted field of fire protection engineering. The purposes of the Society are to advance the art and science of fire protection engineering and its allied fields, to maintain a high ethical standard among its members, and to foster fire protection engineering education. Its worldwide membership includes engineers in private practice, in industry, and in local, regional, and national government, as well as technical members of the insurance industry. Twenty-seven chapters of the Society exist, located in major cities of the United States and in Canada, Australia, Belgium, and England.

Membership in the Society is open to those possessing engineering or physical science qualifications coupled with experience in the field and to those in associated professional fields. There are five major grades of membership: Member, Associate, Affiliate, Technologist, and Student. The grades of Member, Associate, and Affiliate require an engineering degree or comparable physical science education coupled with from one to eight years of engineering experience, several of which must be in fire protection engineering. The recently created Technologist grade requires as a minimum a two-year degree in a fire science/technology curriculum or at least four years' experience in an occupation that has a direct bearing on fire protection matters. The grade of Student is available to individuals with a sincere interest in pursuing a career in fire protection currently enrolled in a curriculum of accepted standing. Students may also organize into student chapters of the SFPE.

Benefits of membership in the SFPE include recognition of professional qualifications by one's peers; chapter participation; receipt of the "Bulletin," a newsletter with regular features; receipt of "Technology Reports"; receipt of the "Yearbook," a biennial directory of members; permitted attendance at the Annual Meeting and Technical Seminars; an insurance plan; an awards program; a public information program; and representation by committees at the national level. The SFPE is a member of the Engineers Joint Council and the Association for Cooperation in Engineering, and has a Liaison Agreement with the National Society of Professional Engineers.

Society Headquarters are at 60 Batterymarch Street, Boston, Massachusetts 02110 (617-482-0686).

The American Insurance Association (AIA) (formerly the National Board of Fire Underwriters [NBFU])

There are many insurance organizations that contribute to our knowledge of fire science, but the unquestionable parent of them all is the American Insurance Association, organized as the National Board of Fire Underwriters in July 1866. Its history of more than one hundred years parallels fire protection developments over that period. Looking at some of the highlights will give us a better understanding of the organization.

Following the Civil War, fire insurance in this country was in a state of chaos compared to the strictly regulated business we know today. Almost anyone could start an insurance company and set the rates at whatever he thought he could collect. This resulted in cutthroat competition and insurance policies that were often valueless.

The New York Underwriters Association circulated a petition to companies outside of New York suggesting the formation of an organization for the common good. While waiting for the results of this inquiry, a conflagration occurred in Portland, Maine, that resulted in many companies going out of business, leaving policyholders unable to collect for their losses. This graphically demonstrated the need for organization. Seventy-five insurance companies signed up and the National Board of Fire Underwriters was born, with these objectives:

1. To establish and maintain, as far as practicable, a system of uniform rates and premiums.
2. To establish and maintain a uniform rate of compensation to agents and brokers.
3. To repress incendiarism and arson by combining in suitable measures for the apprehension, conviction, and punishment of criminals engaged in this nefarious business.
4. To devise and give effect to measures for the protection of our common interests and the promotion of our general prosperity.

The new organization attacked the problem of establishing rates by setting up local boards of agents, providing them with rating formulas, and beginning to collect loss information to establish a statistical base. For a time, this plan worked. The fire insurance business began to prosper again. In fact, there was talk of creating a rating bureau with six departments to cover the United States. The disaster of Portland was soon forgotten, however; agents began cutting rates and companies accepted high risks at low premiums to stay in business. Then, in 1871, came the great Chicago fire, ten times as disastrous as the one in Portland. Once again fire insurance companies took a beating, both financially and in lost public confidence.

Interest in the National Board revived and the insurance business looked to that organization to restore order. By 1874, 90 percent of the fire insurance premiums in the United States went to the stock company members of the National Board. Insurance executives heading up the organization began to take a new look at their operations and began to bring fire prevention into insurance. About this time fire protection was first recognized in fire insurance rating, and a definite change began to take place in the National Board. By 1877, the membership voted to give up control of rates and leave that function to local boards and bureaus. By 1889, a complete rebuilding of the organization was under way, with emphasis on safe building construction and control of fire hazards, water supplies, and fire departments. The National Board of Underwriters was very active in the organization of the National Fire Protection Association and has continued support and cooperation to the present time.

When we look at the 1901 version of the National Board's objectives and compare them with those adopted in 1866, we can readily see the change from self-interest to public service. These objectives, still the guide for the organization, are as follows:

1. To promote harmony, correct practices, and the principles of sound underwriting. To devise and give effect to measures for the protection of the common interests, and the promotion of such laws and regulations as will secure stability and solidarity to capital employed in the business of fire insurance, and protect it against oppressive, unjust, and discriminative legislation.

2. To repress incendiarism and arson by combining in suitable measures for the apprehension, conviction and punishment of criminals guilty of that crime.

3. To gather such statistics and establish such classification of hazards as may be for the interest of members.

4. To secure the adoption of uniform and correct policy forms and clauses and to endeavor to agree upon such rules and regulations in reference to the adjustment of losses as may be desirable and in the interest of all concerned.

5. To influence the introduction of safe and improved methods of building construction, encourage the adoption of fire protection measures, secure efficient organization and equipment of Fire Departments, with adequate and improved water systems, and es-

tablish rules designed to regulate all hazards constituting a menace to the business. Every member shall be bound in honor to cooperate with every other member to accomplish the desired objects and purposes of the board.

The old organization with the new approach was literally tested by fire a few years later: the great Baltimore fire occurred in 1904, followed by the San Francisco earthquake and fire, in 1906, the largest peacetime conflagration in history. Almost all Board members met their financial obligations, although it was a severe strain. Insurance rates all over the United States reflected the enormous losses for several years. But more important, the two conflagrations led to a major change in the organization's approach to public service.

At the Board's first annual meeting after the Baltimore fire, some early fire scientist pointed out that nothing had happened during the fire that a study of conditions could not have foretold. Members decided to make such studies in the larger cities of the country and hired a crew of engineers to make reports. In 1905, a report on San Francisco's fire defense system concluded with this statement: "In fact, San Francisco has violated all underwriting traditions and precedent by not burning up. That it has not done so is largely due to the vigilance of the fire department which cannot be relied upon indefinitely to stave off the inevitable." This report, available to members six months before the disaster, did a great deal to establish faith in the new approach.

Originally, the organization of the National Board centered on the executive committee, composed of top executives of capital stock fire insurance companies. Reporting to the executive committee were separate committees for each activity or department. These included Actuarial, Arson, Engineering, Laws, and Loss Adjustment. The General Manager coordinated all of these activities at the staff level and acted as secretary to all of the committees.

Before World War II, fire insurance companies were generally separate from the casualty and automobile insurance underwriters but, as time passed, more and more companies went into multiple-line underwriting. This led to the consolidation of fire and casualty insurance companies. The eventual effect on the National Board came on January 1, 1965, when the American Insurance Association, the Association of Casualty and Surety Companies, and the National Board of Fire Underwriters consolidated under the name of the American Insurance Association (AIA). This association continued the activities performed by the three predecessor organizations as a trade association serving a large number of companies in the property and casualty insurance fields.

In 1915, the NBFU organized an Actuarial and Statistical Bureau, which kept records on losses sustained by member and subscriber companies as well as on premiums collected. This data was classified into more than 100 occupancies, whether the property was protected or not, whether it was equipped with automatic sprinklers, and the general class of construction.

In November 1964 the NBFU decided that the Actuarial Bureau could better serve the insurance companies as well as the public as a separate organization. Thus, the National Insurance Actuarial and Statistical Association was born. In 1971, this was one of the insurance service organizations that consolidated and formed the Insurance Services Office described below.

The Arson, Theft, and Fraud Department began operating under the NBFU in 1917. The department was headed by a Chief Special Agent and at one time had a force of between 90 and 100 special agents trained to investigate and report on arson fires. They were scattered throughout the states and worked closely with local police and fire officials. Insurance executives recognized the value of this service, but many felt it was the obligation of local law enforcement officials to do the job. As a result, in 1968, this service was discontinued by the AIA.

The objectives of the AIA are to promote the economic, legislative, and public standing of its participating insurance companies. It operates through headquarters in New York and regional offices in San Francisco, Chicago, and Washington, D.C. In 1975, the AIA was servicing 230 subscriber companies. The service best known to fire departments and public officials is the "Municipal Survey and Reports," described in Chapter 10, inherited from the National Board of Fire Underwriters. In 1971, this service was transferred to the Insurance Services Office described later in this chapter.

The Engineering and Safety Service of AIA was not limited to municipal surveys; other important activities still carried out by this association include printing and distribution of more than 160 publications on safety and fire protection subjects, research, and reports on special problems (such as "Fire Hazards of the Plastic Industry"), catastrophe reports, and suggested codes and ordinances such as the "National Building Code," and a "Fire Prevention Code." An especially valuable publication to the fire service is the "Special Interest Bulletins" that are available to all fire protection interests.

Insurance Services Office (ISO)

Prior to 1971 nearly every state and province had its own insurance rating bureau with its own interesting history and organization. They were important contributors to our knowledge of fire protection and were the connecting link between the fire service and the fire insurance companies.

On January 1, 1971, the Insurance Services Office was formed through the consolidation of several national insurance service organizations.

They included the Fire Insurance Research and Actuarial Association, the Inland Marine Insurance Bureau, the Multi-Line Insurance Rating Bureau, and the National Insurance Actuarial and Statistical Association. By January 1, 1972, most of the state and regional fire rating organizations had become a part of ISO and by 1976, there were only eight independent rating bureaus, located in seven states and the District of Columbia. There, the ISO functions as an advisory organization for fire and allied lines of insurance. (See Appendix B for a list of independent fire rating organizations.)

The basic function of a rating bureau is to make and publish advisory rates for fire and allied lines of insurance for its members and subscribers. This operation is designated as "pricing services" in the ISO organization and since the fundamental principles of ratemaking are to analyze the probability of fire starting, of fire spreading, and fire extinguishment, we can see that experts working in this field must have a good working knowledge of fire science.

The Public Protection Grading Services evaluates the fire defenses in most municipalities and fire districts in states in which ISO is licensed as a fire rating organization. The large cities (over 500,000 population) are considered to have enough fire experience to develop their own loss statistics and are not graded under the grading schedule. The inspection and grading of smaller cities has been a function of the rating bureaus since the early 1920s. The details of this public protection operation are considered so important by the authors that an entire chapter (Chapter 10) is devoted to the subject.

The Private Protection Services deal mainly with automatic sprinkler installations and other systems including automatic fire alarm and extinguishing systems. Engineers from this department pass on plans and inspect installations to assure compliance with standards. Special rates are established for property thus protected.

An important advantage of the consolidation of the rating bureaus under ISO is the possibility of uniformity in all rating methods. ISO has given this high priority on the list of things that can be improved. It is developing an improved rating schedule that will provide consistency in ratemaking and will eventually replace some sixteen schedules now in use throughout the United States to establish rates on similar risks.

Underwriters' Laboratories, Inc. (UL)

Electricity was first used for large scale illumination in Chicago at the Columbian Exposition in 1893. The rash of fires resulting from this installation showed local insurance underwriters that electricity would require supervision and control if it were not to become another destructive hazard. They hired William H. Merrill and set up a small laboratory over the stables of the fire salvage company in Chicago. Merrill's work on electrical installations at the Exposition was so productive that he began testing other hazards. Before long, with the assistance of the National Board of Fire Underwriters, Underwriters' Laboratories was founded.

Until 1968, the American Insurance Association sponsored the Underwriters' Laboratories and representatives of capital stock insurance companies made up a large part of the board of trustees of UL. This was in spite of the fact that there has been no cash contribution from the insurance business since 1917. In 1968, the corporate structure of the Underwriters' Laboratories was reorganized, removing it from the control and supervision of the capital stock fire insurance companies. Although the insurance industry is still represented on the UL board, it is no longer the dominant influence. The UL, a nonprofit corporation, is supported entirely by fees paid by manufacturers who want to have their products tested and for follow-up inspection service on listed products. The objectives of the Underwriters' Laboratories are:

> By scientific investigation, study, experiments and tests, to determine the relation of various materials, devices, construction and methods of life, fire and casualty hazards, and to ascertain, define and publish standards, classifications and specifications for materials, devices, construction and methods affecting such hazards and other information tending to reduce and prevent loss of life and property from fire, crime and casualty.
>
> To contract with manufacturers and others for the examination, classification, testing and inspections of buildings, materials, devices and methods with reference to the life, fire and casualty hazards appurtenant thereto or the use thereof; and to report and circulate the results of such examinations, tests, inspections and classifications to insurance organizations, other interested parties and the public by the publication of lists and descriptions of such examined, tested or inspected materials, devices and products, by the provision for the attachment of certificates or labels thereto, or in other manner as from time to time may be deemed advisable.*

To carry out this objective, the UL employs more than 2000 people, including 400 engineers. The organization is divided into seven departments: Burglary Protection and Signaling; Casualty and Chemical Hazards; Electrical; Fire Protection; Heating, Air-Conditioning and Refrigeration; Marine; and Follow-Up Services. The executive offices are in the testing station in Chicago, a site occupied by the UL since 1905. This houses the Electrical Department, a section of the Fire Protection Department, and the headquarters for the Follow-Up Services Department.

The testing station at Northbrook, Illinois, supplements the Chicago

---

*From the UL certificate of incorporation; published in the UL annual report, 1976.

laboratory and provides outdoor space for tests that cannot be conducted safely inside. This station also houses the Burglary and Signaling, Casualty and Chemical Hazards, Heating, Air-Conditioning and Refrigeration Departments, and most of the Fire Protection Department.

In Melville, New York, another testing station has facilities for testing electrical devices and fire alarm signaling equipment for the convenience of East Coast manufacturers; still another, in Santa Clara, California, tests electrical, gas, and oil equipment. The Santa Clara station also tests heating, air-conditioning, and refrigeration equipment and has a Steiner Tunnel for testing the flame spread of interior finishes.

The latest testing station is located at Tampa, Florida. It does some burglary protection and signaling testing as well as testing of some electrical and marine products.

To examine and test more than 40,000 devices annually, the UL has developed and published some 400 internationally recognized safety standards. To obtain the label or listing of Underwriters' Laboratories, a device must meet these minimum standards. If found acceptable, it is listed under the manufacturer's name, along with its catalog number, in one of the seven volumes published annually. Bimonthly supplements keep all lists up to date. A label may be attached to a device, indicating it is listed by the UL, only if the product is subject to "label service," a part of follow-up services. These labels are bought by the manufacturer from Underwriters' Laboratories and affixed to the product. (See Figure 3-3.)

Follow-up services, carried out in various ways, assure continued compliance with UL standards. Reexamination services consist of continued listing of the product and factory visits by a member of the Laboratories' staff (at least once a year) to test samples of the most recent production. If this reexamination indicates that the product is not up to standard, corrections must be made by the manufacturer or listing is discontinued. Under "label service," a similar examination checks the manufacturer's own inspection program. If deficiencies are found, corrections must be made or UL-approved labels removed from the devices. Frequently, labeled products are purchased on the open market and tested in the laboratory as a check on factory inspection work. Some 547 inspectors in 135 American cities and 42 foreign countries carry on this work.

Until the late 1960s Underwriters' Laboratories did not test products for public protection, but limited their testing program to devices that had to do with private protection only. Prior to the time ISO took over the rating bureaus for instance most of these bureaus made acceptance tests on pumpers for cities without charge. ISO eliminated this service and, since then, UL has made rating tests on pumpers and will probably test aerials and elevating platform trucks in the future.

FIGURE 3-3. Some examples of Underwriters' Laboratories labels.
Courtesy of the Underwriters' Laboratories, Inc.

Underwriters' Laboratories contributes to fire science in many other ways. During examination and testing of so many products, information of great importance develops. Much of this has been published in the many publications put out by Underwriters' Laboratories. These include a quarterly magazine, "Lab Data," which contains articles covering UL products evaluation, research developments by UL engineers, and organization information about the Laboratories, as well as articles of general interest involving product safety. Other educational literature is published on such subjects as acoustical noise measurement facilities.

Associated Factory Mutual Insurance Companies (FM)

In 1835, a New England textile mill owner, Zachariah Allen, was dissatisfied with his fire insurance premium and decided to do something about it. His mill was carefully constructed with heavy plank floors and well maintained, yet his insurance rate was no better than the owners of mills with a much higher risk. The concept of rating was foreign to the insurance business at that time so companies turned a deaf ear to Allen's argument for lower premiums. He was impressed with the idea of mutual insurance, as introduced to America by Ben Franklin in 1752, but no mutual company would insure a factory. So Allen called in some of his friends who also owned well-constructed mills. Thus the Manufacturers Mutual Fire Insurance Company, first of the Associated Factory Mutuals, was born in Providence, Rhode Island.

Allen's revolutionary idea was sound: Only manufacturing plants of superior construction and maintenance would be insured by the company, and fire prevention methods would be studied and put into practice as they proved successful. The idea worked so well that business people in other states and manufacturers of other products began forming mutual fire insurance companies. By the early 1900s there were 28 of them, but consolidation has reduced the number to seven today. These firms make up the Associated Factory Mutuals.

From the beginning, insistence on good maintenance and fire protection facilities has made the FM plan a success. Today, extensive inspection by carefully trained engineers assures these conditions in every risk insured.

Probably the greatest contribution made to fire science by FM is its research in methods of loss prevention, work that has been going on for more than 70 years. At first this dealt mainly with the field of hydraulics. The fire service recognizes formulas for computing friction loss in fire hose developed by John R. Freeman of FM more than 50 years ago. Much more extensive work has been done on the development of automatic sprinklers. Building materials and fire protection equipment had to be appraised for the benefit of the insured and this led to the Approved Equipment Service. Like the Underwriters' Laboratories, Factory Mutual labels certain products to indicate its approval: ◁F.M.▷

Loss prevention studies became so important that a corporation was formed, owned by the Factory Mutual Companies, with one objective: "to help policy holders make their properties and production facilities safe from damage by fire, explosion, wind and related hazards." Its name is Factory Mutual Engineering Corporation, and its headquarters are at Norwood, Massachusetts.

Testing and research facilities were maintained in Norwood from 1942 to 1967, when a new research center, on a 1,500-acre tract in West Gloucester, Rhode Island, was dedicated.

Among FM's outstanding accomplishments is the development of a spray head for automatic sprinkler systems that provides more effective distribution of water. The protection of rubber tire storage, high-piled storage, conveyor openings, and fire spread on insulated metal roof decks are a few of the many fire protection problems FM research has helped to solve.

Although the testing facilities of Factory Mutual Laboratories do not cover the variety of devices that Underwriters' Laboratories does and its symbol is not seen as often, the FM label has the same prestige as the UL label.

American Gas Association (AGA)

Other laboratories in the private sector that are of particular interest to the student of fire science are those of the American Gas Association. The AGA maintains two laboratories, one in Cleveland, Ohio, and one in Los Angeles, California. They test and inspect only gas appliances and gas appliance accessories. These devices are submitted to AGA Laboratories for certification as to compliance with national standards.

For the purposes of the certification program, gas equipment is divided into two classes: appliances that are self-contained gas-burning devices such as gas ranges, water heaters, and space heaters; and accessories that are devices used in connection with appliances such as gas valves, thermostats, and other control devices.

Like the other laboratories discussed in this chapter, AGA is a non-profit organization, and the fees charged to manufacturers cover the costs of tests and inspection of the devices. Certification of appliances or accessories is granted only upon compliance of its design with national standards and then only for a limited time. Certification is renewed each year contingent upon satisfactory examination of the product at the manufacturer's plant.

When an appliance or accessory is certified by AGA Laboratories as being in compliance with standards, the manufacturer is entitled to use the AGA certificate symbol on the device.

Industrial Risk Insurers (IRI)

In 1975, Industrial Risk Insurers was formed through a merger of the Factory Insurance Association (FIA) and the Oil Insurance Association.

Industrial Risk Insurers (IRI) insures industrial properties only after an inspection by their engineers indicates that the risk meets IRI's standards regarding fire and safety. Inspections of insured risks are made at regular intervals and reports are prepared covering all areas of loss prevention and control. Reports include information on (1) extension of protection systems to new buildings and processes, (2) application of new protection methods resulting from research, (3) preplanning for emergencies, and (4) proper maintenance of and testing of equipment and administration of procedures.

Inspections are also made of any losses that may occur or for consultation regarding fire protection or safety problems. In other words, IRI insures only especially well protected properties with assurance that the protection will be maintained. The owner is recompensed not only by obtaining insurance at a lower premium rate, but also by receiving expert advice on protection problems.

The headquarters of IRI are in Hartford, Connecticut, with regional offices in Chicago and San Francisco. There are 28 field offices in major cities throughout the United States and more than 500 fire protection engineers serve IRI policyholders.

Underwriters' Laboratories of Canada (ULC)

Underwriters' Laboratories of Canada is a nonprofit organization, incorporated in 1920. It maintains and operates laboratories and a certification service for the examination, testing, and classification of devices, construction, materials, and methods to determine their relation to life, fire, and casualty hazards, and/or their value in the prevention of crime.

ULC operates under the sponsorship of the Canadian Underwriters' Association, whose membership comprises some 140 leading joint stock, fire, and casualty insurance companies in Canada. Underwriters' Laboratories of Canada is a completely separate, Canadian entity, without any financial, legal, or other connection with Underwriters' Laboratories, Inc., in the United States. The two do, however, maintain some technical liaison on matters of mutual interest.

ULC's Scarborough, Ontario, establishment contains 25,000 square feet of office and laboratory space. Laboratory equipment includes a 25-foot tunnel surface for fire hazard classification of building materials, two furnaces for fire resistance tests on floor or roof assemblies, a tower room for tests of factory-built chimneys and Type B vents, an electrical laboratory, and a fully equipped fire service hydraulic laboratory, used for both tests and demonstrations. There is a staff of about 45. In addition, ULC is represented in about 20 locations across Canada as well as in the United States, Britain, and Europe.

From 1950 to 1964, all ULC listings appeared in a single booklet,

"List of Inspected Appliances, Equipment, and Materials," published every year or two, with supplements every six months. Since September 1964, this publication has been divided into two separate booklets by transferring all listings of construction materials to the newly compiled Volume II. The separation provides a convenient reference for architects, engineers, and others principally interested in construction materials and services. It serves as a complementary document to Supplement No. 2 of the National Building Code of Canada.

The ULC Fire Council, organized in 1952, provides advisory technical assistance for ULC's examination and testing projects. The 31 Fire Council members are specially qualified individuals representing federal, provincial, municipal, and insurance inspection authorities across Canada. The Interprovincial Gas Advisory Council gives technical advisory assistance on ULC tests of gas fired equipment.

Like Underwriters' Laboratories, Inc., ULC is completely self-supporting. Revenue for its operations comes from the engineering, listing, and labeling fees charged for product examination, testing, and listing; its labeling and follow-up services are also similar to those of UL.

REVIEW QUESTIONS

1. Name some of the ways in which the National Bureau of Standards contributes to fire science.

2. What is the National Fire Prevention and Control Administration and what are some of its functions?

3. Name four federal organizations whose function is to control fire safety in transportation.

4. Name the fire safety functions of the following federal government organizations:
   (a) National Park Service
   (b) General Services Administration
   (c) Department of Defense
   (d) Department of Housing and Urban Development
   (e) Consumer Product Safety Commission
   (f) Occupational Safety and Health Administration
   (g) National Forest Service

5. Name some of the functions of State Fire Marshal offices that control fire hazards or otherwise contribute to fire science.

6. What are the functions of State Forestry Departments?

7. What are the functions of a State Office of Emergency Services?

8. Describe some of the State Fire Training programs.

9. What are some of the fire protection functions at the county level of government?

10. Outline the functions and operations of the National Fire Protection Association.

11. What are the purposes of the Society of Fire Protection Engineers?

12. What are the functions of the American Insurance Association in promoting better fire protection?

13. The basic function of the Insurance Services Office is to set insurance rates. How does that contribute to fire science?

14. Name the three principal United States laboratories in the private sector that test products for fire safety and describe briefly how they function.

4 Fire Department Organization

In spite of our best efforts to prevent it, fire remains a constant threat to human life and property. Fire science has still not been able to keep every hazard from causing a fire. Indeed, considering modern manufacturing processes, the prevalence of hazardous materials, and the crowded conditions in our cities, it is a tribute to the fire prevention work of our fire departments and engineers that there are not more fires. But when a fire does break out, the fire service is expected to keep it under control and put it out.

Extinguishing fire is the biggest job for any fire department. To do this takes more employees and more expensive equipment than any other aspect of fire department work. The resources and methods now in use for putting out fires as quickly and completely as possible demand such precision in their use that fire fighting fully deserves to be called a science. And, the operation of a modern fire department calls for a disciplined, well organized force of fire fighters and officers.

Semimilitary Organization

As we saw in Chapter 2, the first fire fighting organizations were loosely organized groups of volunteers. This created serious problems because volunteers might or might not show up when needed; teamwork, an essential part of any attack on fire, was often lacking. Clearly, something had to be done to overcome these defects. The obvious answer was a strong form of organization and the strongest in sight was the military.

In the early stages of Western civilization, the military was present at every level, protecting the people and affecting community life and development to a great extent. Most able-bodied males had military experience, making this form of organization familiar, if not attractive. Those who wanted fire protection knew that a paramilitary organization would guarantee teamwork regardless of individual convenience. There would be a group of disciplined workers, constantly on duty, ready to follow directions or orders issued by superior officers. The choice was inevitable, and the form has persisted down to today.

Confusing Terms

Since the U.S. Army was the model for early fire fighting organizations in most American cities, the use of Army terms to describe various parts of our fire departments became the practice. While this vocabulary has largely persisted, there are often great differences in the meaning of the words in the military organization and in the fire service. For example, a fire company rarely exceeds seven persons on duty, including an officer; a standard infantry company (U.S. Army) rarely numbers less than 125 persons and six officers.

Fire departments vary in their form of organization and, unfortunately, also differ in the way they use words to describe parts of the organization. While this can be confusing, these individual variations are not important. What is important is to understand the principles of overall organization and operation. Regardless of what department is joined, if a new member understands these principles the words used as "labels" will soon make sense because they are used to describe a certain function.

With this caution in mind, let us consider the organization of a typical fire department.

Chain of Command

Every effective fire department, regardless of any special makeup, is marked by the quick response of its members to command or instruction. There is no time to debate tactics when a fire is attacking a building or endangering human life. To prevent delay because of difference of opinion, fire department officers are given the same authority as military commanders. This chain of command extends upward as well as downward, through every level of responsibility. Regardless of the units involved, there will always be one officer whose directions become the orders that all those below must follow. And regardless of size, in any recognized department (see Chapter 10) the chain of command always has a chief officer at its top. In a very large department, there may be deputy chiefs, district (assistant) chiefs, and battalion chiefs, holding positions of intermediate command. These officers wear the gold insignia of rank.

Figure 4-1 shows the five crossed trumpets of a chief and the three crossed trumpets of an assistant chief; deputy chiefs rate four crossed trumpets, and battalion chiefs, two. This device reflects a phase in the history of fire fighting when commanding officers, like college cheerleaders, used their voices, amplified and directed by speaking trumpets, to direct their crews. By association, the speaking trumpet came to represent command rank. The color of the badge reflects the position in the organization's hierarchy, with gold for chief officers and silver for other positions.

Chief   Deputy Chief   Assistant Chief

Battalion Chief   Captain's Insignia

Assistant Chief's Badge

FIGURE 4-1. Officers' insignias. Courtesy of V. H. Blackington & Co., Inc.

Constant Readiness

Another reason for the adoption of a paramilitary form of organization is that any fire department must be ready to respond 24 hours a day, seven days a week. The saving of life and property can depend on the promptness of fire fighters and machines. Many calls will be to combat fire or apparent fire threats; some will be other types of emergencies such as drownings, suicide attempts, or electrical accidents; a few will take the form of the trapped pet or small child. Regardless of the reason, a full emergency force must always be available to handle any situation that may arise.

But unlike the military, where individuals may be on the firing line for days at a time with only snatches of rest and time to eat, fire departments meet the need for round-the-clock service by a series of shifts. As discussed earlier, in the early days of organized fire departments fire fighters worked 24 hours a day, every day, with time off to go home for meals. Gradually, departments began to give days off. The 1920s marked the adoption of the two platoon system by the fire service. Under this system, fire departments had two full crews with one crew on duty at a time. This meant an individual workweek of 84 hours. As additional days off were given, the workweek was shortened accordingly. At present, most large fire departments in the United States have three complete crews assigned to each fire company. This system is referred to as the three platoon system.

Methods of rotating and scheduling duty hours for these three platoons vary widely from fire department to fire department. The number of hours required for any shift is greatly influenced by the number of fires fought during that shift and the total number of duty hours in the platoon's normal workweek. Some cities work 8-hour shifts; others have a shift system divided into day shifts of 10 hours and night shifts of 14 hours. Most fire departments, however, have adopted a 24-hour duty cycle.

Where an entire department is on a volunteer basis, alarms, direct phone lines, and similar alerting devices connected to the members' homes and businesses serve to maintain the 24-hour service standard. Small departments, with a skeleton staff of paid members, often have a standby group of volunteers or individuals who are paid for duty by the call. Again, the full complement is summoned to duty by alerting devices, and fairly good results are obtained. While many volunteer and "combination" departments bear a large part of the burden of fire protection and suppression in our communities, they suffer from one major handicap: because of the time that must elapse before arrival "on site" or at the station, response is necessarily slower than for fully staffed, paid departments. This can prove costly, because delay may allow a fire that could have been extinguished easily in its initial stages to gain headway and become a really large fire with heavy losses.

Coming back to paid fire departments, members cannot be expected to stay awake and alert 24 hours at a time. Therefore, the standard fire house provides a place for them to cook meals, exercise, study, and sleep. The fire house is a home away from home and usually has the same conveniences found in today's average home.

Relaxation between alarms can be a very important consideration, particularly after a large fire or during inclement weather. In addition, many fire departments find that regular physical fitness activities are important, tending to reduce accidents, increase physical stamina and strength, and keep the minds of those engaged active and alert.

Medical history has begun to show the economic value of regular physical examinations, diet control, and programmed physical activities in spaces especially provided and equipped. The lost duty time involved is insignificant, as is the cost, when compared to the possible full loss of one or more members due to accidents or illness caused or contributed to by exhaustion, obesity, or physical disability such as a heart attack or back or respiratory problems.

There are many times when a series of alarms will keep the on-duty shift from getting any real rest, but there is usually some time to grab a fast meal. If this begins to occur continually, another work schedule will usually be established, such as three 8-hour duty shifts.

Regardless of which shift is worked, the people who live together, who train and face emergency situations together, tend to develop a comradeship similar to that found among members in the armed forces. The longer they work together, depending upon each other as part of a team, the stronger the bond and sense of shared accomplishment.

Departmental Organization

For members to fight fire effectively, they must be organized. The basic unit in fire department organization is the fire company and the basic member is the line fire fighter. The type of fire company and the number of members a given company needs to perform its function depend on function and apparatus. Ideally, good organization will allow a department and its companies to perform effectively with a minimum number of members.

In practice, the number assigned to a given company depends on the company's function and the size of the entire organization. In turn, the size of any given fire department depends on the ability of a city, municipality, or district to support it. For example, a pumper company in a large department may have as many as six or seven members assigned to duty, while in a small community there might be only two or three members to operate the same apparatus. This difference depends partly upon the kind of area a company protects and partly upon the number of gallons of water per minute required to protect that area or to effectively fight a fire within it. (See Chapter 10.)

As a result of these differences in crew size, ability to lay hose, to connect the pumper to a water supply, and to supply water under pressure for one or more standard 946 liters per minute, lpm (250 gallons per minute, gpm) fire streams is often used as a gauge of the effectiveness of a pumper company. Considering these functions, we can readily see that a two- or three-member company will be less efficient than one with a larger crew. Most chiefs agree that five is the minimum crew size for successful pumper company operations and that companies serving in a high-value or hazardous district should have larger crews.

Ladder companies, normally referred to as truck companies, carry a variety of ladders, forcible entry tools, salvage, and overhaul equipment, plus a large assortment of other tools and equipment. The load is quite varied because ladder companies are responsible for rescue, ventilation, forcible entry, salvage, overhaul, and a multitude of other fire fighting and emergency duties. (See Chapter 7, Fire Fighting Tactics and Strategy, for details.)

A modern ladder truck is equipped with an aerial ladder that can be raised, rotated, extended, and positioned by a combination of cable and hydraulic systems, driven by a power take-off from the truck's engine. Once an aerial ladder truck is stabilized, by chocking its wheels and setting the outriggers, it can easily be controlled by one man. An aerial truck also carries a minimum 50 m (163 feet) of ground ladders, varying in length and type, that are used to carry out some of the aspects of fire fighting just mentioned as the responsibility of ladder companies.

Much has been written about how to get a portable ground ladder off the truck and in place against the side of a building. Every department has its own approved procedure. What is important is that carrying a ladder and positioning it at a fire is a slow and dangerous process; it also requires much more crew work than we might expect. With larger ladders, for example, four to six crew members are needed. When in use, the number that can safely use a ladder is severely restricted by safety requirements. From this, we can easily see that at any fire where rescue demands are heavy or where ladders must be used to advance hose lines, crew size demands will be higher than for pumper operation.

Another type of apparatus that has gained prominence in the fire service is the elevating, or aerial, platform. This is a hydraulically driven device that can raise, extend, and position a basket at any selected point on the upper portion of a building. One strategic use is as a mobile elevated platform for use with heavy stream appliances. (For other uses, see Chapter 5, Fire Department Equipment.)

Company Makeup

Within the average fire company, the leader is generally known as a captain. This varies from state to state and includes such titles as lieutenant or, in some southern states, sergeant. While the name and even the insignia of rank may differ from fire department to fire department, the essence of the job is keeping a group working as a coordinated team, acting as the unit commander of a vehicle and its crew.

An individual selected to lead a fire company is generally one with considerable fire fighting experience and proven ability to lead. Reaching this position is not easy. Career services require applicants to pass written examinations before they can be considered for officer rank; appointment is usually restricted to those making the highest grades. Competition is keen. Today, many first-line supervisors in the fire service are members who have completed college and specialized fire science courses. Many captains have a background of study that includes such subjects as supervision, fire department administration, record and report writing, principles of fire prevention, chemistry, radiological defense, and advanced fire tactics and strategy.

Insignia, designating rank, are displayed on the badge and cap devices of fire fighters. In addition, officers usually wear collar ornaments and have distinctive helmets so that their command rank can be easily ascertained while on the fireground. A captain's insignia of rank ordinarily consists of two silver speaking trumpets, crossed or vertically side by side. The rank of lieutenant is usually shown by a single silver trumpet. Figure 4-2 illustrates these insignia and some worn by line fire fighters.

Captain's Insignia   Lieutenant's Insignia

Driver's Cap Piece   Inspector's Badge

Fire Fighter's Badge

FIGURE 4-2. Insignias of company officers and crew. Courtesy of V. H. Blackington & Co., Inc.

Next in importance in a fire company is the driver of the apparatus. In the days of horse-drawn fire fighting apparatus, driving was dramatic and demanding; today, it may be less dramatic but the volume of traffic and the higher speed of motor vehicles make the job still more exacting. Traffic tends to pile up and become congested around the scene of an emergency. Yet to be a good driver, one must be able to work a large, heavy vehicle through snarled and jammed traffic conditions and park it where it can be put to best use. The driver must also be able to handle the vehicle under adverse conditions such as snow, ice, heavy rain, and damaged pavement and over difficult terrain.

In addition, every time a piece of apparatus leaves its quarters, whether for a drill, inspection, or in response to a call, the value of the apparatus and the equipment it carries are at stake. Modern apparatus is very costly and prices continue to rise. Some pieces cost $100,000; the average cost of an unequipped pumper is about $50,000. When the value of the individuals riding the vehicle, the lives of citizens on the streets when the apparatus is making its run, and the lives and property of the people who turned in the alarm are added to the value of the apparatus and its equipment, the sum of a driver's responsibilities becomes very great.

As if these were not enough responsibilities, the driver is often the person who selects the response route to a fire, decides where to spot the apparatus, and operates the vehicle while at the fire. A practice followed by most modern fire departments is to have the captain pick the route of travel and select where and when to use the apparatus on reaching the fireground. This is preferable because the captain often has the experience, training, and time to plan the action during the response. The captain also has the use of a radio to coordinate the actions of the company with those of other companies arriving at the fire scene or already there. The driver should be left free to give full attention to driving through traffic and operating the apparatus after arrival.

Selecting drivers is obviously a matter of great concern to any fire department, large or small. While all the members of a fire fighting unit should be trained and qualified to assume any position in the company, including that of driver, some may be better suited to the driver job than others. Bear in mind, however, that in some departments driver positions carry special ratings and only those who have previously qualified may be assigned to this duty. Close screening of potential drivers is a must and should extend to their personal habits, attitudes, aptitudes, and mental limitations. A well-run fire department will have a detailed selection program and extensive training facilities.

Although the job descriptions and requirements for drivers vary from organization to organization, they generally include the following elements:

- Good vision, particularly good night vision
- Stable personality

- Mechanical aptitude
- Ability to make intelligent decisions rapidly
- Considerable fire service experience
- Ability to solve mathematical problems, either mentally or with slide rule or hand-held calculator
- Strength and skill in handling heavy apparatus

The need for this range of abilities can be readily understood in the light of what drivers are required to do. Drivers of ladder trucks must be able to position their apparatus where the ladder can be used to greatest advantage; drivers of pumpers must be able to calculate pump and nozzle pressures rapidly, determine the needs of sprinkler and stand-pipe systems, and solve relay pumping problems. Drivers are also often required to perform preventive maintenance on apparatus; in smaller communities, they may even be called on to make major repairs or modifications of the apparatus. In some departments, drivers are responsible for teaching new members in their companies how to drive and how to care for the apparatus. Candidates for driver positions should complete class work and driver training before they are allowed to handle apparatus; those already assigned as drivers are subject to checkout and retraining, when needed.

Titles of the persons assigned to driving and performing many related mechanical jobs vary widely. Some departments designate those who drive as engineers; others use such terms as chauffeur, auto fire fighter, driver-operators, and variations and combinations of one or more of these terms. In line with this, some departments adopt distinctive insignia for their drivers in place of the usual fire fighters' badges. Other departments make no distinction at all for driver positions but simply expect fire fighters to assume these extra duties and responsibilities.

While some fire departments use drivers as relief shift company commanders, this practice is subject to debate. On the positive side, drivers have demonstrated they are worthy of trust and are often willing to assume the added responsibilities of command. On the negative side, it is extremely difficult to operate efficiently as both unit commander and driver-operator, for the reasons mentioned earlier in this chapter.

The basic member of a fire company is the line fire fighter. Although modern fire department companies center around their major pieces of apparatus, the crew assigned to a given company are normally trained to perform many fire fighting functions and are not narrowly specialized. The reason for broad training is that modern apparatus can be used for more than one fire fighting function. For example, ladder trucks may carry a small amount of hose; some even have a small fire pump. Pumpers often carry an assortment of ladders, and in smaller departments, every piece of apparatus carries equipment for limited rescue and salvage work. It is true that a member's specific assignment to a tactical unit may call for specialization, but in general, any fire fighter in a modern fire department is expected to possess the full range of fire fighting skills.

Under the direct supervision of their company officers, these are the people who fight fire, inspect buildings, render first aid, rescue people and animals, take care of the fire apparatus, and train to improve their skills. They are, in a sense, the backbone of the organization and, as such, give the organization its power or render it ineffective. This is doubly true of those who make up the fire service in small communities, whether all volunteers, partly paid, or a combination of the two.

Operational Organization

As we will see in Chapter 10, the Standard Schedule for Grading Cities and Towns of the United States with Reference to Their Fire Defenses and Physical Condition determines the ideal size of any given fire department. To do this, grading engineers use three factors: population of the area to be protected, required fire flow in gallons per minute, and response distances. These factors determine how many companies of each type are needed for the best possible protection. Unfortunately, because of lack of members, money, or public support, the Standard's requirements are frequently not met and a given fire department is smaller than it should be.

Regardless of size, however, any fire department must be well organized to be an effective fire fighting force. The form such organization takes varies from very simple to very complicated. In general, the larger the department, the more complex its organization will be. To analyze and discuss the differences in organization of several fire departments of different sizes would take many thousands of words. Fortunately, management specialists have provided us with a great spacesaver: the organization chart. Such a chart, when properly drawn, clearly shows the relationship between the parts of an organization and the overall structure as well. With this in mind, let us consider the typical operational organization of several different types of fire departments.

Volunteer and Small Fire Departments   Up to company level, most fire departments are quite similar in the manner in which they allocate responsibilities and in the rank that various positions carry. Above this level, many differences exist. The simplest organization, however, is found in a volunteer fire department with one piece of apparatus. In such a department, there is a chief, an assistant chief, and volunteer fire fighters. Occasionally there will be a paid driver.

The solid lines in Figure 4-3 illustrate the operational organization of a larger volunteer fire department with several pieces of apparatus.

Notice how similar this organization is to that for a small fire department, as shown in Figure 4-4, and how both differ from the organization of a public safety department which includes both fire and police protection under a single chief. Figure 4-5 illustrates the organization of a public safety department.

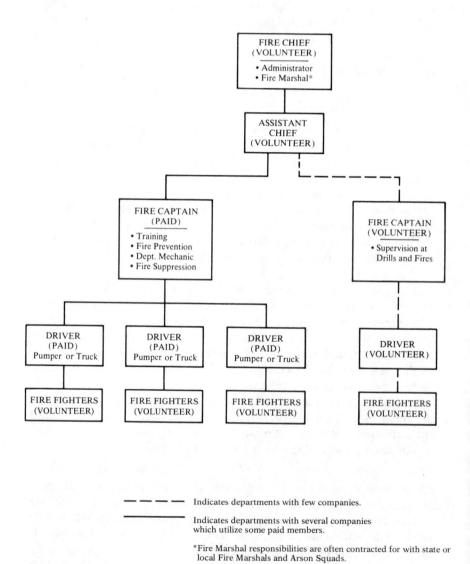

FIRE CHIEF
(VOLUNTEER)
• Administrator
• Fire Marshal*

ASSISTANT
CHIEF
(VOLUNTEER)

FIRE CAPTAIN
(PAID)
• Training
• Fire Prevention
• Dept. Mechanic
• Fire Suppression

FIRE CAPTAIN
(VOLUNTEER)
• Supervision at
Drills and Fires

DRIVER
(PAID)
Pumper or Truck

DRIVER
(PAID)
Pumper or Truck

DRIVER
(PAID)
Pumper or Truck

DRIVER
(VOLUNTEER)

FIRE FIGHTERS
(VOLUNTEER)

FIRE FIGHTERS
(VOLUNTEER)

FIRE FIGHTERS
(VOLUNTEER)

FIRE FIGHTERS
(VOLUNTEER)

— — — — Indicates departments with few companies.

———— Indicates departments with several companies which utilize some paid members.

*Fire Marshal responsibilities are often contracted for with state or local Fire Marshals and Arson Squads.

FIGURE 4-3. Organization of a volunteer fire department.

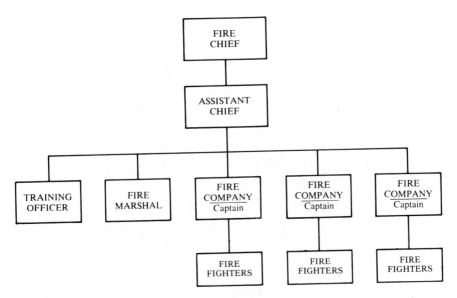

FIGURE 4-4. Organization of a small fire department (less than six companies).

Large Fire Departments   When a department becomes large enough to have a considerable number of pieces of apparatus, and particularly when there are several fire stations spread throughout the protected area, there must be intermediate levels of command and rank to coordinate and conduct the more complex activities of the department. Intermediate command takes several forms. The most common type is based on the number of companies in the department. In this form, a battalion chief commands a battalion composed of approximately eight companies. This permits the companies to operate as a unified force under a single director. The span of control that exceeds eight fire companies at the scene of an emergency becomes unmanageable.

Note: "Span of control" as used herein refers to the effective control one supervisor can expect to have over a specific number of persons and pieces of equipment at the scene of an emergency.

Where there is more than one battalion, the chain of command is commonly extended by creating districts, each headed by a district (assistant) chief. They are the immediate superiors of the battalion chiefs in their district. In very large departments, there may be another overall command position, deputy chief. At the top, of course, there is always the chief of the department. Figure 4-6 illustrates the organization of a very large department.

Another form of intermediate command is based on the shift system common to most fire departments. Each of the two or more teams that

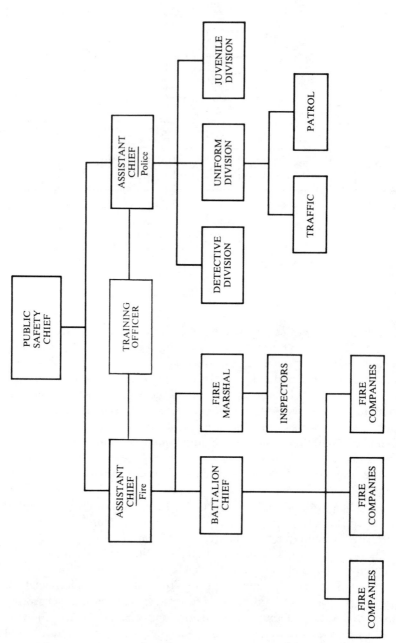

FIGURE 4-5. Organization of a public safety department.

alternate on duty is under the command of its own officers. In turn, these officers may be supervised by a battalion chief or chiefs, subordinate to the chain shown in Figure 4-6.

Functional Organization

There is another aspect of organization: administration and management of the department's affairs. In cities like New York, Chicago, and San Francisco, to name a few, fire departments employ large numbers of people and have budgets that run into millions of dollars yearly. This is big business and running an organization of this size, like any other business, calls for good administration.

Most fire departments in the United States, however, have a headquarters staff that consists of the chief alone, backed up by one or more assistant chiefs or deputy chiefs so that there is always a commanding officer on duty, or the headquarters and staff consist of the chief and a small number of clerical and shop employees. Figures 4-3 and 4-4 illustrate simple organization; Figure 4-6 shows a more complicated organization, including some of the functions we will discuss a little later in this chapter.

One immediate effect of size is that the chief becomes more and more separated from fire fighting operations, more and more the fire department's chief administrator. As a result, headquarters work must be divided in such a way that the chief retains good administrative control without having to give daily attention to all staff functions. If properly done, staff organization spreads the administrative workload so that this is possible. While the operational chain of command remains the same, from chief on down through assistant chief, division chief, battalion chief, company captains, and lieutenants, the number of those reporting directly to the chief will be reduced to not more than five and, better yet, no more than four. Figure 4-7 shows the functional organization of a fire department.

A large fire department must have expert management. Choosing the individuals who occupy the managerial positions on the headquarters staff calls for great care. The primary requisite is managerial skill, but in a fire department, particularly at high levels, technical competence is equally desirable. Such people are not easy to find and, unfortunately, chiefs that are good managers are even more scarce. Since there is little room at the top, few persons who enter the fire service try to prepare themselves to manage a fire department—the job that is the modern fire chief's principal function. This lack of good managers is even worse in volunteer fire departments where chiefs tend to come and go at relatively short intervals.

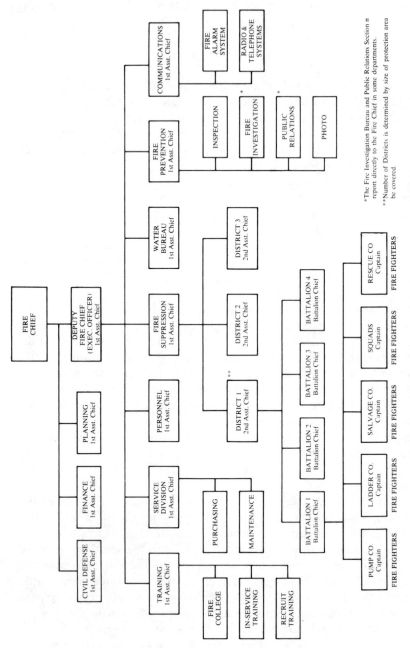

FIGURE 4-6. Organization of a large fire department.

*The Fire Investigation Bureau and Public Relations Section report directly to the Fire Chief in some departments.

**Number of Districts is determined by size of protection area be covered.

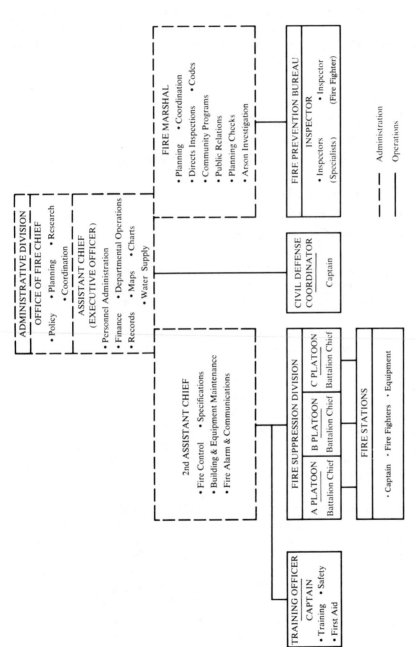

FIGURE 4-7. Functional organization chart of a fire department.

An example of how ill-prepared many chiefs are for the tasks of management can be found in the answers a large group of chiefs gave to the question: What is your principal concern as chief of a fire department? These are some of the responses:

- Getting equipment
- Rating subordinates
- Getting members to drill
- Promoting members to officer grade
- Pacifying the fire commissioners (or mayor, city council, city manager)
- Writing the rule book
- Recruiting members
- Preparing the fire department budget

There were many more responses of this kind. The point is that almost none of these chiefs put administration first and, equally important, few seemed to realize that what they had made their principal concern was something that could and should have been handled by a member of their staff. In other words, these chiefs were concentrating their efforts on one or two aspects of management instead of spending their time supervising and coordinating the entire operation.

Now, what do we mean by management? Management is the art of getting people to do things. The function of management is to create a unified, balanced, refined tool that, when applied in the right way, will accomplish any desired objective without undue strain. Authority is the key to managerial success and delegation of authority is the key to organization. Delegation of authority can be defined as vesting in a subordinate a portion of a superior's authority, while retaining overall control.

Ability or willingness to delegate authority is the mark of a good manager but it can be overdone. Excessive delegation weakens overall control and may even result in stripping the manager of all authority. The grant of authority should have its limits. The better defined these limits are, the more comfortable a subordinate may feel in making decisions and performing his/her duties. Specific delegation of authority, in writing, is the best form, but this can become too confining. If authority is delegated in this way, a channel must be kept open that will allow for change, redefinition, and discussion of the limits imposed.

The position of executive officer (see Figure 4-6) is a good example of delegation of authority. The person who holds this post deals with most of the staff and the members of the line organizations. The executive officer must have strong qualities of leadership and administrative ability as the principal officer in charge of administration. To do the job effectively, the executive officer must have very broad powers and exercise them fully.

Figure 4-6 also illustrates a headquarters staff for a very large fire department. When organized in this way, there is an officer responsible for every aspect of administration and operations plus overall management control, extending from the chief on down. The executive officer bears a heavier burden than the chief. Reporting to the executive officer, up through the chain of command, are those responsible for all fire fighting operations.

Let us consider what some of these headquarters groups do to make the department function smoothly and well. At the same time, remember that the operation of any fire department embraces these same activities; the difference is that in smaller departments these functions are handled by one or two individuals, perhaps with the help of a few clerical and shop employees.

Finance   The fire chief, as the administrative head of the fire department, has the full responsibility of developing the financial budget for the department. Every expenditure must be justified and once the budget is approved, the chief has the responsibility to function within its parameters.

The chief may, depending upon the department's size, have a finance officer. However, the responsibility is the fire chief's.

Each governmental body has developed a fiscal policy, budget procedures, and so forth, and the fire chief must work within this framework.

Many organizations have been moving toward the program budget concept. This technique develops goals and objectives and attaches resources for their attainment. A form of measurement can be applied that will permit the chief to determine the value of the program, its cost to the community, and whether or not the program meets the criteria established.

This technique often sheds light on old programs that have been carried out every year but are no longer effective in the community or, in some cases, are more costly than they are worth. It is an opportunity to look at programs and their costs in a more professional manner so that good decisions can be made.

The usual practice today is to have an assigned account number for the fire department with subdivisions for payroll, stations and ground, apparatus and equipment, fire hose, fuel, and housekeeping supplies. With proper clerical help (including accountants where budgets permit), a good finance officer or division head can give the chief precise information on the amount spent from each account, the amount remaining, and the best way to apportion fire department funds. The department should also keep detailed cost records on all major pieces of apparatus and equipment, should budget for their replacement, and should manage to replace part of the fire hose every year.

Fiscal officers should also discourage the practice of juggling the department's funds by diverting budgeted amounts from one category to another. A particularly bad practice is using payroll money to purchase apparatus, hose, or other supplies, and compensating for its diversion by reducing company strength. Most municipalities have a bonded city finance officer directly responsible to the governing body to handle the finances of all city departments.

Public Relations   A large part of the public relations work of any fire department is educating the public to understand the danger of fire and to cooperate in keeping safety at a high level. Failure in this effort can mean the difference between a good and a bad fire record. It is not enough to have a fire prevention code; the citizens of a city must be convinced that code provisions are reasonable and necessary for the public safety. Experience proves that it is far better to try to persuade people to accept regulation than it is to try to force regulation upon them—especially when enforcement means extra expense. Since many of the provisions in our present fire prevention codes call for costly construction features or special equipment, persuasion becomes doubly important. The alternative is argument, resistance, and probably lengthy court battles.

The public relations section of a department must conduct a continuous well-thought-out program of education. Personal visits by high-ranking officers to large plants, mercantile establishments, and other "target hazard" buildings (buildings potentially able to cause a fire that will result in a large loss of life or property) will reach small groups and individual owners. Wider acceptance of the fire department's plan and ideas comes from talks before civic, fraternal, and service groups, backed with well-written material. Both speakers and writers should be particularly careful to keep technical terms to a minimum.

Special campaigns call for public relations work. Usually, the key to success is support by local newspapers, television, and radio. An example of such a campaign is Fire Prevention Week, the week in which October 9th, the anniversary of the great Chicago conflagration, falls. Proclamations by the president, governors, and mayors call attention to the waste fire causes. An alert public relations section backs this up with newspaper stories, television appearances, and school talks. In many cities, open house at the fire stations helps build goodwill for the department and shows the public how important the fire department is to its safety. Spring Cleanup Week and voluntary dwelling house inspection campaigns also call for this kind of departmental public relations activity.

One of the most fruitful activities a public relations section can engage in is the support of local fire prevention committees. In fact, fire departments are the sponsors of a majority of these organizations. Supplying the committee with information and materials encourages support of all departmental activities and safety goals.

The public relations section also has the problem of protecting the department against unfair criticism and seeing that correct reports of any fire are carried by the various news media. Very often, a department is severely criticized immediately after a bad fire. This can be checked, or even prevented, if the newspapers, radio, and television are shown how aggressive, intelligent fire fighting reduced potential fire loss in spite of existing unsafe practices or dangerous construction. Similarly, as a background to any account of a fire, the public relations section should provide details of attempts to correct fire hazards that were made before the event. When fire fighting activity damages property or a building, a public relations person should take pains to explain to the owner or tenant why such action was necessary.

Above all, an alert public relations section will not miss the opportunity to press the department's recommendations for improvement in safety practices on individuals and the entire community immediately after a large fire.

Because many types of fire occur only at certain times during the year, the public relations section can concentrate its efforts on stressing specific hazards for that time. Depending on geographical location, in fall and winter the target is heating hazards; in spring, rubbish and brush; and in summer, lightning and spontaneous ignition in farm produce. School fire safety programs, geared to the age of the audience, encourage safe practices and fire prevention in homes.

We speak of a "public relations section," but we actually mean everyone in the organization, because all are a part of this activity. Departments large and small have a need to get the message out to the people. Therefore, there must be a commitment by every member. Fire department members must always be alert to the fact that what they do and say, how they do it and what they look like doing it, is the department's public image. It is important to realize that this is true whether on or off duty.

Fire Investigation

The investigation of fires for their causes and the detecting and apprehending of arsonists is correctly the function of the fire department.

Fire investigation is by its nature the basis for fire prevention. Only an in-depth analysis of what sequences of events enabled a fire to start, enabled it to spread, and how and where it was controlled (e.g., fire fighting, structural design, lack of fuel) can help prevent future fires. Additionally, fire investigation includes the observations of everyone involved, and at fires themselves there are many fire fighters who will be able to shed light on the nature of the fire, its progress, and so forth.

Arson and the malicious activities of many persons cause a significant economic impact each year in this country and abroad. There are many motives that contribute to these crimes. Large fire departments have fire investigation bureaus, staffed by specialists and a special arson squad. Smaller departments may have one or more trained specialists or may utilize other neighboring department staffs. We will discuss the arson squad first.

Actually, the legal authority for investigating fires and prosecuting people who have "arranged" for a fire is usually given to the state fire marshal (see Chapter 2). Even where this is the case, local fire department chiefs are commonly given the rank of deputy fire marshals, or the state marshal has the authority to deputize them. As a result, most chiefs are responsible for investigation of all fires that occur in their locale, particularly fires that appear suspicious. This responsibility is delegated to the fire investigation bureau (or section or division). (*Note*: In some departments, the fire investigation section and arson squad are part of the fire prevention bureau. Regardless of the organization chart, their responsibilities will be the same.)

Every fire is initially presumed to be accidental in origin. To establish the crime of arson, the state must prove that the building or structure (even an automobile) was burned deliberately, by someone who knew what he was doing when he started the fire, or by someone else with whom he conspired to set the fire. During the notorious outbreak of arson in New York City during the 1930s, fires were set by professional criminals, called "torches," for a fixed price or a percentage of the expected insurance settlement. Owners of the buildings and businesses were held equally responsible for the acts of those whom they had hired.

While the number of incendiary (deliberately set) fires is a small proportion of the total in most cities, a large part of those that show a high loss may be laid to fires that were deliberately set. Most cities do not make adequate investigations. When fires are thoroughly inves-.tigated, incendiarism turns out to be an important cause.

In practice, an arson investigator does not speed to the scene of every fire. Even if it were desirable to do so, this would be impossible in a city of any size. Action depends on what the fire fighters and officers at the fire scene notice and report. The presence of a number of fires in different places; trails of flammable material, particularly gasoline and kerosene in areas where they are not normally used; a candle stub in a box packed with excelsior or scraps of film—all point to arson. Another suspicious circumstance is buildings or premises that have been cleared of fixtures, furnishings, or stock before the fire broke out. In the absence of these almost-certain indicators, the district chief will decide whether further investigation of an ordinary fire is desirable. When the fire is a large one, however, the arson squad will be on hand to start work at once. Another instance where the arson squad might respond to the first alarm is where the records show that the owner or proprietor has a record of frequent fires. Serious and fatal injuries, as well as explosions, are also investigated thoroughly.

On the scene, the arson squad looks for such evidences of incendiarism as mentioned above, fingerprints on articles found near the fire and on nearby doors and windows, forcible entry marks other than those caused by the fire department—anything unusual or unnatural. Heating and electrical systems are examined to make sure the fire was not started by any defect in either system; equipment such as stoves, heaters, and ovens comes in for the same scrutiny. If the fire fighters noticed anything suspicious, such as an extremely rapid rate of burning, the smell of fuel oil (where there should be none), smoke of a distinctive color, a person running away as they approached, or open doors and windows (when they should be closed), the arson investigator will have their statements taken promptly by a staff stenographer. Any evidence will be carefully photographed before it is moved and will be subjected to chemical analysis, if needed. A draftsperson should make accurate floor-plans and sketches of the fire scene.

Recent court decisions indicate that on-the-scene investigation may take place during a time period which is "reasonable" for the fire department to complete their operations. After this time, investigators must obtain court warrants to enter private property. This serves to protect the constitutional rights of property owners, and should not be detrimental to adequate fire investigations.

Back in the office, arson investigators can add to what they have learned at the scene from fire department records of past fires. Good practice will guarantee a file of all fires investigated, filed by date of occurrence. This file will show the building, the fire, location, owner's name, occupant's name, and loss sustained. The investigation section may also have a card file listing persons who have had suspicious fires in the past and another showing suspicious fires by location. Some departments also use a checklist that includes such material as dates when insurance was written, amount outstanding (if possible); how long the property was owned; financial status of the owner of a business; results of interviews with suspects, suspects' employers, and witnesses; and data from interviews with the fire fighters who fought the fire.

By combining all sources of information, an arson investigator can arrive at a fairly sound conclusion as to whether a fire was incendiary.

Proving a case of arson against one or more persons is another matter entirely, and may not be possible no matter how convinced the investigator may be of guilt. The choice of whether to prosecute is up to the district attorney's office as part of the criminal justice system.

Fire investigation work follows much the same pattern in attempting to discover the cause of a fire. In addition, fire investigation prepares a detailed inventory of the damage done by the fire that will show exactly how things stood when the department left the scene. Through agreement with insurance companies, the department also receives reports of losses in small fires where the department was not called and, of major importance, a report of all large losses paid. By preparing accurate plans and drawings, showing the circumstances of the fire, the investigation bureau can help point up necessary charges in the fire prevention code, the building code, and so forth, and the precise departure from safety practices that caused the fire to start. Photos are particularly helpful in showing point of origin, direction of spread, interference with protective devices such as fire doors, or failure of sprinkler systems.

A copy of all fire investigation reports is customarily sent to the office of the state fire marshal. Where the fire is considered suspicious or there is evidence of arson, this report must be made. Many departments also supply the NFPA with copies of reports on all major fires and on any fire where the loss exceeded $100,000. A few cities supply this type of information to local and special agents of fire insurance companies. Based on its record, the fire investigation section can supply the chief with a reasonably accurate figure of annual fire loss in the city.

Planning   No manager can operate efficiently without planning. Good planning saves time, conserves people, money, and other resources, secures stability and adaptability, controls and diminishes the severity of an emergency, gains respect and confidence from the members of the department and the public, and decreases internal tensions and criticisms. A good planning section helps the chief select the department's objectives and develops the policies, procedures, and programs needed to reach these objectives.

Actually, Figure 4-6 should show a link between the planning section and every other group, bureau, or section. The reason becomes obvious when we consider the most common goals of planning: adding and replacing equipment and apparatus, developing staff, encouraging staff growth and establishing promotional policies, improving and conducting inspections, determining the scope and content of training programs needed, and revising and updating salary schedules.

Planning depends on accurate data. To get this data, the planning officer must have access to good records, kept up to date. He must also be able to summarize these records for use in making decisions. The biggest problem is the gap between formulating a plan and getting it put into effect. This is particularly true for operational planning.

A solution that has had considerable success is group planning, where those concerned with what is under consideration are called in and asked their opinions and suggestions. Plans that are the result of this type of thinking generally get more support from those who will carry them out. There is also less chance for serious error since the final plan embodies a broader base of experience and knowledge than any one individual possesses.

A good planning section is always willing to look at what other departments have done or can do, to listen to the theoreticians within its own department, to look for the broader objective, and to invite criticism and suggestion by those who are informed about the subject.

Personnel   A good fire department costs money, and the largest part of this money is spent on salaries for the people who provide fire protection. When a person is recruited and trained, he or she represents a considerable investment of departmental funds. If the new member proves unfit for the fire service or leaves to take another job, this money is lost. A good personnel section will prevent or reduce the chances of this turnover in a number of ways.

Most fire department hiring and firing must meet local, state, and federal guidelines. As a result, the first way that a personnel section can cut excessive turnover is by working closely with its civil service counterparts to establish correct standards and qualifications for applicants seeking to enter the fire service. This same cooperative approach is also needed to formulate accurate job descriptions at every level.

Next, and perhaps equally important, a good personnel section evaluates new employees carefully before a probationer is given permanent appointment. Another very valuable service is conducting exit interviews with persons who leave the fire service voluntarily. Their reasons for leaving frequently offer a good indication of needed organizational changes or reveal hidden problems at company or platoon level that need correction.

The personnel section is where every member's service record is kept. Here are the papers that show performance ratings, medical record, leave record (both sick and annual), and training record. This is also the place for records that show when an employee is eligible to take a promotion examination, standing on the civil service list, length of service, eligibility for retirement, or qualification for a disability pension. See Appendix D, for examples of evaluation forms.

When well run, a personnel section can be an invaluable aid to a department's chief officers and all its members. The information in its files relates to practically every aspect of fire department administration and operations. A badly run section, however, is worse than useless, since improperly or inadequately kept records can cause all sorts of difficulties for everyone concerned. It is up to the executive officer to see to it that the personnel section is run well and has enough people in it to handle the detailed and accurate records a fire department must have to function at its best.

Training   Training, depending on the size of the department, may be handled by a section of the headquarters staff or it may require a division, headed by an assistant chief in charge of training, with a number of other training officers. Whatever the size of the group, it should provide the executive officer and the chief with highly detailed plans for departmental training. It should also coordinate and direct instruction at all levels, with a particular eye on what is taking place at battalion and company training sessions. When needed, a good training section will not hesitate to bring in outside specialists or agencies either to conduct instruction or to evaluate the training itself. This is also the group primarily responsible for developing the training program for all units and all types of department members under a chief's control.

In a large municipal fire department, the training division may supervise or help conduct operational training, including the use of new types of apparatus. It may offer courses for officers in more advanced and technical subjects, train instructors for work at company levels, train fire fighters and officers to prepare for civil service promotion examinations, and conduct specialized training, both technical and fire-related.

Crossing administrative boundaries, a training section may handle training in fire prevention work, rescue, and first aid that qualifies members for advanced courses; it may conduct courses for executive officers and potential executive officers in management and administration. Similarly, the training section may give the planning section help by holding conferences that appraise the results of using new fire fighting techniques, discuss special problems, and reexamine command procedures. These conferences can also benefit officers in charge of fire fighting operations.

Whatever advanced training may be offered will necessarily be limited by available facilities. In smaller cities and departments, the training section may provide only basic training in fire fighting skills and techniques. State colleges help bridge the gap when they offer annual or short courses on various fire service subjects. For fire departments that are too small to have a qualified instructor, there are often state college or university extension programs that cover elementary and basic fire fighting skills and techniques.

Fire Prevention Bureau   Practically every city with more than 50,000 population has one or more full-time fire inspectors; most cities of over 100,000 have a separate staff division in their fire department to handle this work. Even smaller towns need at least one fire prevention inspector who is familiar with building construction as well as fire science. This is because fire prevention is even more important than extinguishing fires.

The head of a fire prevention bureau should be one of the best officers in the department, a good executive, able to manage and organize the work well. He or she should have adequate office space and clerical help, transportation, and enough full-time inspectors attached to the bureau to do the job. These inspectors should be specially qualified technically and skilled in dealing with the public. Never forget that a good fire prevention bureau will get the reputation of being the best place to get the answers to all questions about fire prevention. If this flow of information is cultivated, it can save many dollars in fire losses and greatly reduce the danger of loss of life.

While the fire prevention bureau should coordinate all the fire prevention activities within a fire department, bureau inspectors are primarily concerned with enforcing fire prevention codes, making technical surveys, and inspecting target hazard properties. Appendix D, pages 327 and 329, are examples of target hazard inspection forms.

The importance of having a fire prevention code and having it enforced properly is reflected in the Standard Schedule (see Chapter 10) where 13 percent of the deficiency points used in grading a city's fire defenses are determined on this basis. A well-thought-out fire code will cover such subjects as: storage of explosives and ammunition; storage and handling of flammable liquids; placement of lumberyards; installation of dry cleaning establishments; safety features for garages; handling and use of paints, varnishes, and lacquers; use of acetylene and compressed gases; storage and sale of pyroxylin plastics (nitrocellulose); handling, purchase, and distribution of nitrocellulose X-ray, photographic, and motion picture film; use, storage, and processing of combustible metals; limitations on type and use of fireworks; safety measures for fumigation operations; regulation of storage and handling of matches; storage of hazardous materials; safety measures for industries that produce potentially explosive dusts or vapors; installation and safety measures for refrigeration equipment; amount, number, and maintenance of fire exits in buildings; general maintenance standards for buildings; and safety measures for maintenance of vacant lots, dumps, rubbish, and waste.

Quite a list. But in practice, life is made easier for the bureau's inspectors by the requirement that most hazardous installations and occupations, such as welding or blasting, have licensed operators and permits for installation. Written and practical examinations determine whether an individual can get a license; before a permit is issued, the bureau's inspectors go over the premises and check for compliance with the code's provisions. If changes are needed to meet code standards, they must be made before the permit is issued.

In issuing permits for hazardous occupancies, the work of the fire prevention bureau and the building department overlap considerably. Few individuals, if any, are equipped to handle the vast range of subjects covered by the fire prevention code and the building code. As a result, specialists must pool their knowledge and judgment, making cooperation essential if disputes are to be avoided. In general, the building department has the last say on structural features, but this is subject to fire department approval of certain features such as venting, thickness of fire walls, and so forth.

In a well-run fire department, target hazards such as key industries or property where life hazard is high come in for special attention from the bureau's inspectors. Typical target properties are institutions, such as hospitals, schools, and nursing homes; places of public assembly, such as theaters, churches, and bowling alleys; places where flammable liquids or gases are stored in bulk; large stores and warehouses, particularly shopping centers; chemical and petroleum plants, including plastic manufacture; lumberyards and woodworking plants; marine depots; and any place that uses or stores reactive or toxic materials.

Constantly changing techniques and materials used in buildings, manufacturing, and processing create new hazards and new problems in fire prevention. Each industrial building in a city should be carefully inspected to learn its structural, occupancy, or exposure hazards. Groups of buildings should be studied as a larger unit in order to estimate conflagration dangers. An industrial city with closely-built areas of manufacturing plants and densely populated districts, or a concentration of high-value buildings surrounded by areas with combustible buildings, make fire prevention work difficult.

Inspectors who visit plants must be particularly well qualified technically. If they are not, their advice will be ignored and any suggestion they make resented and opposed; if they are, they can often persuade the owner or manager to adopt most proposed changes. These inspections should also include a discussion with management of preventive measures and cooperative action. (See Chapters 8 and 9.) Plans for cooperative fire fighting are usually written out and then tested by an exercise or drill.

Duplicates of the bureau inspectors' reports and findings on plants and target hazards are often given to the fire company or they are carefully incorporated in the preplanning records for use in future operations against fire. Conferences, following drills or testing of prefire planning, help to work out the role of the individual company in the event a fire does occur. (See Chapter 7 for discussion of preplanning.)

Although inspecting target properties can make up the biggest part of the work load of the bureau's inspectors, they frequently get an assist in this work from insurance company teams and similar groups of engineers and inspectors from the parent company of a large local installation. Smaller properties, because of their high risk, may also get attention from the insurance company but they usually lack good private protection systems. Public buildings and institutions, however, must depend on either the fire prevention bureau or state agencies to protect them against fire hazards.

The bulk of all fire prevention work in any department is done by routine fire inspections, conducted by members of individual fire companies, under the supervision of district chief officers. One method used for this type of inspection is to have the members of a radio-equipped company make a block-by-block inspection of prescribed areas. The company stays in service to answer fire calls. The other method is to assign members to make detailed inspections, usually on foot, within their protection district. This has the disadvantage of taking people away from the station and reducing the force available to fight a fire. Appendix D has representative inspection forms.

Regardless of the method used, routine inspections serve to accomplish several things:

- They see that exits and exit routes are open and free from obstacles or interference.
- They reveal the existence of common fire hazards that might otherwise go undetected.
- They familiarize the fire fighter with individual properties where a fire might occur, providing a sound basis for prefire planning.
- They provide a community with the realization that the fire department will check up to see that regulations are complied with.

The value of active company inspection is well established. Intensive fire prevention raises the level of fire protection and greatly reduces both the number of fires and the loss of life. A recent development is scheduling company inspections through the fire prevention bureau with as much as half the department's companies out inspecting at one time.

Company inspectors also take care of "small unit" properties and dwelling inspections. Small unit properties are such occupancies as plumbing, electrical repair, carpentry, upholstery, and paint shops, one-story stores, and small mercantile buildings. Where gasoline is stored in underground tanks, neighborhood filling stations are small units. These units should get attention at least quarterly and as often as monthly when inspections are scheduled properly. Where violations or poor housekeeping show up, inspections are made more frequently.

Records of small unit inspection are kept at local fire stations as part of the preplanning file. What information they contain depends on instructions and departmental policies. (See Appendix D for forms.)

The greatest impetus to dwelling inspections comes from the fact that about two-thirds of all persons who are killed annually by fire die in residence fires. Further, the number of dwelling fires average over 550,000 annually and cause financial loss estimated at more than $325 million. But these inspections can be made only as a service to the householder. Getting cooperation requires an extensive educational campaign. A special gain from dwelling inspections, if done properly, is a public feeling that the fire department is concerned with the personal safety and property of everyone in the community.

Actual inspection is usually carried out by two people from a local company, working in the area the company protects. The householder accompanies them while they point out oridnary simple fire hazards, such as trash in cellars and attics, piles of old newspapers and magazines, unsafe storage of gasoline or similar flammable liquids, overloaded electrical circuits, and so forth. A common side effect of this inspection campaign is a frenzied cleanup before the inspectors can arrive. Seemingly, people do not like to show how badly they keep house. Some jurisdictions also show that women when alone at home will accept female inspectors more readily than male inspectors, perhaps as a result of high crime-to-the-person statistics in an area. This factor should be taken into account to increase the acceptance of this type of inspection.

Since effective company inspection depends on fire fighters and officers, they must be trained in fire prevention. This means crossing boundaries once again. The training section should offer elementary instruction to every member of the department, using such texts as the NFPA Inspection Manual. Such a course should cover the more common fire hazards and defects in fire protection; as the legal authority, the inspector has to have hazards removed or corrected and specific instructions on how to handle the problems that arise most often. Even after this training, an experienced inspector or company officer should go with persons making their first inspection. Advanced training should be offered fire fighters who wish to become members of the fire prevention bureau. After experience at actual fires, people with a good technical education are particularly well suited for this work.

No matter how well trained, every inspector should carry a building record form. These forms (Appendix D) serve as a checklist on construction features, fire protection equipment, and hazards of the building.

They also provide an immediate record of fires in a given building and a signed record of inspection.

There is some debate as to whether inspectors should carry detailed checklists when making an inspection. Some departments insist upon this; others argue it creates unnecessary paperwork. (Appendix E contains representative checklists.) The best answer seems to be that all beginners probably need a checklist; most experienced inspectors do not, and where an inspector is covering a large number of places, with only moderate hazards, a checklist form can be very helpful and speed the work greatly.

Violations that constitute an immediate hazard to life and property should be corrected at once; the inspector should not leave the premises until the danger has been eliminated. Examples of such violations are locked exit doors, passageways blocked or partially blocked with stored materials or goods, automobiles parked under fire escape ladders, and open containers of low flash point flammable liquids. When a fire prevention inspection reveals a minor hazard, the inspector should try to get the owner or tenant to correct it without any formal action. The hazard should be noted on the inspection report, however, and the inspector should later check to see if anything has been done. Clear-cut violations of the fire laws, such as rubbish accumulations, call for a formal notice that tells the owner what is wrong, how it can be corrected most economically, and how long a time is allowed to make the correction. After the stated time is up, a reinspection follows. (See Figure 4-8.) If no action has been taken, a second notice of violation advises that the owner will be subject to fine or imprisonment or both if the condition persists. When the time stipulated in this second notice runs out with nothing done, court action follows. Usually, the fire prevention bureau handles formal violations and all inspectors are required to turn over notices of this type to it. The bureau handles the violation until it is corrected or the matter is concluded.

Communications   "Communications" may be a bureau, a division, or a section, depending on the size of the department in question, or it may be a separate governmental department providing this service. Its work embraces the operation and maintenance of the alarm, radio, and communications systems of the department. The communications group should be headed by a technically qualified superintendent who reports directly to the executive officer or the chief. Members of the group should also have the necessary technical training and experience to service the communications equipment and make repairs and installations whenever needed. To serve its purpose, this section needs an adequate force of dispatchers and two or three shifts of crews, each headed by a supervisor, who are able to operate the equipment, dispatch

DEPARTMENT OF FIRE

OCCUPANCY TYPE				MOUNTAIN FIRE DISTRICT		DATE	
IBM CODE				**BRUSH CLEARANCE NOTICE**			
MAP BOOK	PAGE	PARCEL	TRACT	LOT	BLOCK	OCCUPANCY PHONE	

TO		STREET	CITY	STATE	ZIP CODE
ATTENTION		LOCATION OF HAZARD			

Provide and maintain firebreaks on the above described property in conformance with Section 57.25.20 of the Municipal Code.

Section 57.25.20 requires that any person who owns, leases or controls any land in the Mountain Fire District shall maintain firebreaks upon such land in accordance with the following:

1. All native brush situated within one hundred (100) feet of any structure, regardless of whether said structure is located upon such land or upon adjacent land shall be maintained at a height of not more than eighteen (18) inches above the ground nor less than three (3) inches above the ground and shall have removed therefrom all branching growth and leafy foliage.

2. All native brush situated within ten (10) feet of the outer edge or edges of the usable road surface of any highway, street, alley or driveway serving more than one residence shall be maintained at a height of not more than eighteen (18) inches above the ground nor less than three (3) inches above the ground and shall have removed therefrom all branching growth and leafy foliage.

Approved specimen native shrubs may be retained, provided they do not exceed two hundred sixteen (216) cubic feet in apparent volume, and provided they are spaced from other specimen native shrubs, native brush or structures at a distance of not less than eighteen (18) feet. A specimen native shrub is defined as an individual shrub which is within the definition of 'Native Brush' and which is trimmed up one-third (1/3) of its height or two (2) feet above the ground, whichever is less, and from the vicinity of which has been removed all dead wood, duff, and combustible litter, and which is not among those plants identified as 'Extremely Hazardous Native Brush'.

Approved plants which are identified as 'Fire Resistive Plants' do not fall within the definition of 'Native Brush' and may be used without restriction.

This notice does not require the removal of trees, nor shall it be considered as authority for performing any work on adjacent property without the written permission of the owner of such property. If the requirements stipulated above extend beyond your property line boundary, it is then required that you comply with these requirements to your property line.

LEGALLY DISPOSE OF ALL VEGETATION AND DEBRIS WHICH IS REMOVED.

THIS IS A FINAL NOTIFICATION

FAILURE ON YOUR PART TO COMPLY WITH THIS NOTICE BEFORE ....
UNLESS ARRANGEMENTS ARE MADE OTHERWISE WITH THE UNDERSIGNED PRIOR TO THIS DATE, WILL SUBJECT YOU TO PENALTIES PRESCRIBED BY SAID ORDINANCE OR LAW, OR CAN RESULT IN THE AFOREMENTIONED VEGETATION AND GROWTH BEING REMOVED BY CITY AUTHORITIES, IN WHICH CASE THE COST OF SUCH REMOVAL SHALL BE ASSESSED UPON THE LOTS AND LANDS FROM WHICH SUCH VEGETATION IS REMOVED AND SUCH COST SHALL CONSTITUTE A LIEN UPON SUCH LOTS OR LANDS UNTIL PAID.

FOR ADDITIONAL INFORMATION PHONE	BY ORDER OF THE CHIEF ENGINEER & GENERAL MANAGER
	By _____
	Inspector Assignment

FIGURE 4-8. Example of a violation notice.

apparatus as needed, maintain the alarm system, and service radio and telephone communications. Sometimes, all municipal signal systems are under a single electrical department. Even when this is true, there should be a fire department alarm section, because every chief must keep control of this system and know its condition at all times.

Water Bureau   As we will see later, the Standard Grading Schedule considers water supply the most important element in rating a city's fire defenses. As a result, the fire department must work closely with the city water department on all fire protection plans. The location and condition of the city's fire hydrants and the amount and reliability of the water flow they can supply are vital to fire fighting operations. The fire department also has a direct interest in how the city supplies water to private protection systems (see Chapters 8 and 9) and the kind and amount of water flow private systems are able to develop for themselves.

In addition, fire department water planning includes standby sources such as ponds, streams, harbors, or any source that can be used when a catastrophe, such as an earthquake, destroys municipal supply systems. In this area, the fire department develops potential sources by itself, constructing concrete ramps or pads close to water for pumper access, special storage tanks, and connections and pipelines for relaying emergency water supplies.

To handle the proper development of water supplies and to correct any weaknesses discovered in the system, larger fire departments have their own water bureau, headed by a water officer. This officer should be familiar with water supply systems but should look to waterworks engineers for expert opinion on design and operation of systems.

Because the fire department has no control over the municipal water department, the water officer and staff can only make recommendations for desired changes or improvements. Ordinarily, given the necessary information the city water department will modify its construction plans, particularly extensions of the distribution system, and adjust its operations to meet fire fighting needs.

A good example of how operations are tailored to fire needs is the common arrangement that lets the water officer at the site of a major fire call for a shutoff of part of a city system to give area hydrants more water at greater pressure. Similarly, the water officer can call for an increase in city main pressure by increased pumping.

The fire department water bureau should have a complete file on the public water supply system. This would include maps showing the location and size of mains, gates and shutoff valves, location and capacities of hydrants, and location and potential flow from auxiliary water supplies. These maps should also indicate areas where flow or pressure is low enough to call for special pumping operations. The bureau should also have similar maps and information about private protection facilities. Members of the water bureau itself do not have to gather all this information by themselves. To do so in a city of any size would call for a large staff. Instead, most of the information is available from the city water department and from water-resource maps prepared by each fire company for its first-due area. The bureau's staff analyzes and summarizes this information and supplements it where necessary by personal inspection.

The major function of the water bureau is to set up a program of investigations, inspections, and tests that will give day-to-day information on the water available for fire protection. The water officer and the staff should keep a close watch for extensive repairs, extensions, or other factors affecting water flow in the municipal system. They should also be responsible for investigations and inspections to determine that water is available and its source is reliable. This includes actual testing of facilities for flow, pressure, and performance.

Inspection of hydrants, private water supplies, and auxiliary water supplies is assigned to fire companies in the district protected. Regular written reports of inspection and test results (see Figure 4-9) are forwarded to the department's water bureau and used as the basis for any corrective action that should be taken. These reports also provide an up-to-date record of municipal fire protection capabilities.

Members of the water bureau staff frequently carry the load of inspecting fire protection systems that have no connection with the municipal system or are far outside its immediate service area. These systems are usually based on tanks, cisterns, or reservoirs or they may use platforms, ramps, or floats to mount pumps and carry piping to supply the system. The bureau should conduct tests, using fire department pumpers to see if such systems perform adequately when installed, and should inspect them at regular intervals to be certain that they are kept in good working condition.

Supervision of hydrant installation is another task frequently assigned to the water bureau of the fire department. Having a bureau member on hand to advise on placement can prevent the serious handicap of having to back and turn fire apparatus in narrow streets or alleys, ensure the placement of outlets at the proper height above final grade, and save department maintenance costs by seeing that the hydrant base is properly drained, particularly where freezing may be a problem. The accessibility of all gate valves in the mains that must be open for proper hydrant service can be checked at the same time. Sometimes installation crews or paving operations cover these valve boxes. This can be corrected immediately when a member of the water bureau is on hand. Otherwise,

FIRE DEPARTMENT

HYDRANT RECORD

NUMBER _____ DATE INSTALLED_____

TYPE   72, 74, 75, 76, 78 SIZE MAIN _____

LOCATION _____ _____

FLOW TESTS

DATE	STATIC	RESIDUAL	PITOT	GPM	GPM @ 20	REMARKS

FIGURE 4-9. Example of a hydrant inspection record.

it may take a major fire to discover this serious defect in the water system.

Service (Supply) Division   Because the scope of the activities of the service division in a large fire department is so broad, we will limit our discussion to a brief summary of its principal functions.

The purchasing section of the service division is responsible for buying everything a department needs and uses, from the soap and paper goods used in individual fire stations to the newest, largest, fully equipped pumper. In most cities, by law, all major purchases must be made on the basis of competitive bids, and in some cities even relatively minor items are subject to this requirement.

While a purchasing section may prepare its own specifications for any item, the accepted practice is to use NFPA standard specifications, particularly when buying apparatus, fire hose, and fire fighting equipment. These specifications eliminate costly gadgets and frills and almost guarantee fair competitive bidding. Special provisions allow for modifications to meet particular local requirements.

Where a city has a central purchasing agency, it is up to the purchasing section of the fire department's service division to prepare the technical specifications for bids and to advise the agency on such matters as relative quality of products and reputation of manufacturers.

The service division is responsible for servicing all equipment and apparatus. The department's repair shops are part of this division and the maintenance program, carried out by individual fire companies, should be prepared and supervised by the service division staff. As part of this work, the service division maintains records that show the condition of each piece of apparatus, dates of purchase, major repairs, semiannual overhaul, annual pumper test, and depreciation. Summaries of these records are used by the finance and planning sections to budget for replacement and new purchases.

Some service divisions also handle the work involved in testing hose, running pumper tests, checking extinguishers, repairing respiratory equipment, tools, and specialized electrical equipment. A few cities with well-equipped shops and skilled workers manage to build or rebuild major pieces of apparatus at considerably less cost than buying new apparatus.

Buildings require daily care but they also need painting and repair from time to time or even structural change or addition. The service division normally takes care of housekeeping chores at department headquarters and training schools and facilities; station care is left to the individual companies. Major structural repairs or changes are usually handled by contractors on bid, but regular building maintenance work, including painting, is commonly handled by the service division.

Handling the flow of supplies and equipment within the fire department is also up to the service division, usually by a section that handles all company requisitions. Nothing may be issued from stores without a requisition. This section develops and administers the procedure for making out requisitions, having them approved, and seeing that they are filled without undue delay. The same group sets up and supervises the system of stock control (how much of a given item is on hand), keeps a perpetual inventory record, and conducts physical inventories of stores as needed. These accounting measures guard against theft, waste, or needless accumulation of supplies. The record of use and purchase forms the basis for estimating and planning purchases during the year to come.

Another recent development is a section that handles purchase and issue of personal equipment to members of the fire service. Departments customarily provide their members with fire fighting clothing, including helmet, boots, waterproof trousers, turnout coat, gloves, goggles, and special shoes. There is a growing trend toward departmental purchase and supply of all clothing, except underwear and socks, including dress uniforms and overcoats and several sets of fatigues. The clothing section prepares the specifications for these articles and, because of quantity purchases, is able to get quality goods at lower prices than the individual members could. Departments that follow this practice need not provide a uniform allowance and can be certain that their members will present a neat, well dressed appearance.

Records

We come now to the bugaboo of all fire fighters, chiefs, battalion chiefs, and captains: paperwork. As we have just seen in our discussion of the organization of a large department, most of its sections or divisions rely very heavily on records. In this respect, a fire department is no different from any other large organization. The point is that any information that is recorded should be needed and used. Too many departments call for routine reports that are only filed away and forgotten. This wastes both time and money.

There can be no question, however, that three types of records must be kept if a department is to have any real record system at all: the company record book, or journal; the company diary, or day book; and fire reports.

Company Record Book, or Journal

This book contains a permanent record of company activities and equipment. If kept properly, it represents a good way for the company officer to keep control of the organization. A typical company record book contains such personal information as:

- Roster of crew assigned
- Leave for sickness or injury
- Vacation time
- Personal efficiency ratings
- Personal fire fighting equipment issued
- Individual training records:
  Schools attended
  Courses completed
  Drills on apparatus, equipment, radio equipment, and station
  Group exercises, including more than one company
  Personal information related to each member (depending on policy)

In its records and reports section, the company record book carries these items:

- General orders to the department
- Reports from the chief
- Fire report
- Copies of fire inspection reports
- Prefire planning
- Water resources maps, showing location of sprinklers, standpipes, hydrants, mains, and valves in the district; also auxiliary sources
- Hydrant location and servicing record
- Apparatus assigned
- Equipment assigned
- Amount of hose assigned
- Station equipment
- Repairs to apparatus, equipment, hose, and station
- Record of gas, oil, and lubricants used
- Laundry service record
- Personal clothing issued

The record must be constantly updated, by daily attention and entries, and kept free from useless material by culling out unnecessary or outdated records and reports. How long material must be kept is generally a legal question, but once the answer is obtained, this should guide the review of any files. Some departments have extensive file storage just to keep company records from reaching unwieldy bulk.

Company Diary, or Day Book

The purpose of this record is to show each day's activities, hour by hour. For those familiar with the military, part of the entries cover the same ground as the "morning report": members present or absent and why, remarks on the condition of equipment and apparatus, any action taken by the company officers, the comings and goings of members of the company, people present from headquarters or other levels of the department. In addition, there are brief summaries of all calls and operations, including running time and time spent fire fighting. The officer in command checks the entries and signs, if approved. When this book is filled, it should be kept as a permanent record because it clearly shows what a company has done over a period of time.

In large fire departments, the company diary is used to prepare a monthly summary of company work. In small departments, this summary usually need be made only quarterly. The value of this consolidation is that it clearly shows the chief, executive officer, and planning section where the most activity occurred and, presumably, where the largest work force is needed. It may also show a need for relocating a fire station.

Fire Reports

The report after a fire may be either simple or complex, and it may be completed by the officer-in-charge (OIC) of a single company or by the OIC for many companies. Previously, manually prepared fire incident reports were written by a variety of people differing in language skills. The description of a fire and its cause varied with the skill of the author or the importance placed upon that part of the report. This made it difficult, if not impossible, to comprehend what actually took place and to compare one alarm with another. Figures 4-10 through 4-12 are examples of some of these manually generated reports.

These manual reports often are very subjective. Historically there was no standard terminology that would assure that two fire causes would be described in the same way, or would be interpreted later by another person in the same way as the reporter meant it to be. This, of course, made it very difficult to generate accurate statistics that would enable researchers to identify what needed to be done to prevent recurrences.

The National Fire Protection Association (NFPA) started developing a fire reporting system in 1938. Progress continued over the years and, in 1969, the uniform language contained in NFPA pamphlet 901, Uniform Coding for Fire Protection, was developed.

FIRE DEPARTMENT
REPORT OF FIRE

TIME:_____   ALARM NO:_____

DAY:_____   DATE:_____

LOCATION:_____   NAME OF BUSINESS:_____

HOW REC'D:_____   INFORMATION REPORTED:_____   TYPE:_____

APPARATUS RESPONSE:	SECOND ALARM	THIRD ALARM	GENERAL ALARM
FIRST ALARM	TIME:_____	TIME:_____	TIME:_____
_____IN_____	_____IN_____	_____IN_____	_____IN_____
_____IN_____	_____IN_____	_____IN_____	_____IN_____
_____IN_____	_____IN_____	_____IN_____	_____IN_____
_____IN_____	_____IN_____	_____IN_____	_____IN_____
_____IN_____			

OUT ON ARRIVAL?_____ HOW EXTINGUISHED:_____

HOSE USED ▶   3" $2\frac{1}{2}$" $1\frac{1}{2}$"   BOOSTER   TOTAL   GALS.   APR. TANKS:_____

NO. LINES: _____ _____ _____ _____ _____   WATER   HYDRANTS:_____

NO. FT: _____ _____ _____ _____ _____   USED: TOTAL:_____

LADDERS RAISED ▶   NO._____TOTAL FEET_____ SALVAGE COVERS:_____ BREATHING EQUIP:_____

EQUIPMENT USED ▶ _____

SPRINKLER SYSTEM ▶   USED_____ NO. HEADS_____ RESTORED BY:_____

STANDPIPES ▶   WET_____ DRY_____ RESTORED BY:_____

LOSS OF LIFE:_____ INJURIES:_____

CHIEF OFFICERS PRESENT:_____ _____ _____ _____

FIRE PERSONNEL PRESENT: ON DUTY_____ OFF DUTY_____ F.P.B._____ TOTAL_____

POLICE OFFICER:_____ MUTUAL AID DEPTS._____

SUMMARY OF OPERATIONS & DETAILS OF FIRE:

ORIGINAL - BUSINESS OFFICE   (REPORTING OFFICER)

FIGURE 4-10. Example of "Report of Fire."

FIRE DEPARTMENT

ALARM INVESTIGATION REPORT

TIME:_____
DAY:_____
LOCATION:_____
OCCUPANCY:_____
PROPERTY OWNER(S):_____ ADDRESS:_____
TENANT(S):_____ ADDRESS:_____
BLDG. OR BUS. MGR.:_____ ADDRESS:_____
NOTIFIED: TENANT_____ OWNER_____ UTILITIES_____ TELEPHONE CO._____ BLDG. DEPT._____ ELECT. DEPT_____ OTHER_____

ALARM NO._____
DATE:_____
OCCUPANCY GROUP:_____

ALARM DETAILS:
DISCV'D. BY:_____
REPORTED BY:_____
POINT OF ORIGIN:_____
MATERIAL IGNITED:_____
EXTENT OF SPREAD:_____
CAUSE:_____
LAWS VIOLATED:_____
PICTURES BY:_____ ROLL NO._____
BLDG. CONST.: TYPE_____ HT_____ WALL FRMG_____ ROOF FRMG_____ EXT. WALLS_____
ROOF CVRG._____ INT. WALLS_____ CEILING_____

INSURANCE:
BLDG. INS. BY CO./AGENT:_____
ADDRESS:_____ PHONE_____
CONT. INS. BY. CO./AGENT:_____
ADDRESS:_____ PHONE_____

LOSS INFORMATION:

	ASS'D. VALUATION	INS. CARRIED	ESTIMATED LOSS	INS. CARRIED	INSURED LOSS	UNINS. LOSS	ACTUAL LOSS
1. BUILDING:							
2. CONTENTS:							
3. OTHER:							
4. TOTALS:							

5. EXPLANATION LINE 3:_____
6. EXPLANATION UNINSURED LOSS:_____

DETAILS OF INVESTIGATION:

_____ (INVESTIGATED BY)

ORIGINATOR - BUSINESS OFFICE

FIGURE 4-11. Example of "Alarm Investigation Report."

FIRE DEPARTMENT

ALARM OTHER THAN FIRE

TIME:_____

DAY:_____

ALARM NO:_____

DATE:_____

NATURE OF
LOCATION:_____ EMERGENCY:_____

HOW   INFORMATION
REC'D:_____REPORTED: _____ TYPE:_____

                                                                 HOSE   WATER
APPARATUS RESPONSE:_____ _____ _____ _____ USED:_____ USED:_____

APPARATUS TIME IN:_____ _____ _____ _____

EQUIPMENT USED:_____

NAME:_____ AGE:_____ SEX:_____ PHONE:_____

ADDRESS:_____ CARE GIVEN:_____

RELATIVE NOTIFIED:_____ RELATIONSHIP:_____

DOCTOR CALLED:_____ PRESENT:_____ PHONE:_____

COMMUNICABLE DISEASE (TYPE):_____ MEMBERS EXPOSED:_____

AMBULANCE COMPANY:_____ PHONE:_____

CHIEF OFFICERS PRESENT_____ _____ _____ _____

FIRE PERSONNEL PRESENT: ON DUTY _____ OFF DUTY_____ F.P.B._____ TOTAL:_____

POLICE OFFICER:_____ OTHER CITY SERVICE:_____

DETAILS OF THE ALARM:

ORIGINAL - BUSINESS OFFICE

(REPORTING OFFICER)

FIGURE 4-12. Example of ''Alarm Other Than Fire'' report.

The NFPA pamphlet 901 AM, Fire Reporting Field Incident Manual, has a series of simply prepared, manually generated reports, that utilize this uniform language. Figure 4-13, "Basic Field Incident Report," is an example of the type of form and data used.

Jurisdictions that have access to data processing systems utilize the NFPA 901 Code as their basic reporting language as well. Figure 4-14 is an example of a fire incident report generated by fire departments in California. This system, called the California Fire Incident Reporting System (CFIRS), which became required by the state legislature in 1972 for the purpose of reporting fire problems in the State of California, provides that small departments without a computer send in manually prepared incident reports to the state fire marshal, who will have the data extracted and entered into the state's computer system. Larger fire departments having a computer are permitted to send the data to the state's computer system on tapes, which permits them to utilize their own computerized fire reporting system. Data is compiled by the state and is reported annually to both the legislature and the local jurisdiction who submitted the data.

Figure 4-15 is the reverse size of a CFIRS report; it permits the local fire department to collect data that the state does not require.

The fire problem in the United States increases significantly each year. Attempts to identify the specifics of the problem failed in the past because reliable statistics were not available. Some states had no uniform method of gathering and reporting data, while others had a confusing variety of methods. At best, the data available were fractured; sometimes the data were totally unreliable.

The federal government, therefore, entered into several grant agreements for the purpose of developing a uniform method of reporting fire statistics that would enable fire protection researchers access to valid data. Some of these grants were very detailed and included the total fire management concept, while others concentrated on the collection of fire incident data.

The NFPA received one such grant and established what is now known as the Uniform Fire Incident Reporting System (UFIRS). Figure 4-16 is an example of a UFIRS field incident report used by one West Coast fire department. The reverse of this form, Figure 4-17, enables the OIC to write down the story and diagram anything the OIC feels is significant to the incident.

You will note the similarity in all of these forms—whether a manual form or a computerized data processing form, they use the NFPA 901 Uniform Code as a basic language. Thus, state and national records can be generated with a common language and data can be expected to be reliable and usable by researchers.

BASIC FIELD INCIDENT REPORT

Use words on this Report

National Uniform Fire Reporting System, NFPA Form 901F

COMPLETE ON ALL INCIDENTS

A	Incident No. / Time / Day of Week / Month / Day / Year
B	Correct Address / Telephone No. / Room or Apt. No. / Census Tract
C	Occupant / Owner
D	Address of Owner / Telephone No.
E	Method of Alarm From Public / Type of Situation Found
F	Type of Action Taken / Eng. Co. (1st-In Dist.) / Shift / No. Alarms / Mutual Aid Yes No

COMPLETE IF CASUALTY OR FIRE

G	Fixed Property Use Classification / Property Type / Complex
H	Mobile Property Classification / If Mobile / Year / Make / Model / License No.
I	Area of Origin / Interior Finish / Dimensions / Occupant / Floor Level or Height
J	Number Injured* — Civilians / Fire Service — Number Killed* — Civilians / Fire Service
K	Number Fire Service Personnel Used at Scene — Officers / Men — No. Eng. Cos. Used / No. Truck Cos. Used / No. Other Vehicles Used / No. Other Vehicles Not Used

COMPLETE IF FIRE

L	Equipment Involved in Ignition (if any) / Form of Heat of Ignition
M	If Equipment Involved in Ignition / Year / Make / Model / Serial No. / Voltage (if any)
N	Type of Material Ignited / Form of Material Ignited
O	Act or Omission / Extent of Flame Damage
P	Extent of Other Damage / Flame Spread Factor — (if any)
Q	Smoke Spread Factor (if any) / Estimated Property Loss
R	Method of Extinguishment / Number Streams ¾-1″ / 1½″ / 2½″ / Over 2½″ / Feet of Ladders

*List name, age, sex and description of injury for each casualty on Form 901G. Initialed Endorsements

S — Officer in Charge (Name, Position, Assignment) — Member Making Report (If Different from Above) / Date

COMPLETE ON ALL

☐ Check Box if Remarks are Made on Reverse Side

What Factors Helped Limit Casualties or Fire Spread — Enclosures, Fire Doors, Fire Extinguishing System, Good Access, Good Housekeeping, etc.

What Factors Extended Casualties or Fire Spread: Poor Hydrant Spacing, Lack of Fire Extinguishing System, High Piled Stock — Improper Flammable Liquid Storage, Lack of Cutoffs, Scant Water Supply, Non-Fire-Stopped Walls, Open Stairs, Open Shaft, Weather, etc.

COMPLETE IF CASUALTY OR LOSS OVER $10,000

This Form is for use with NFPA 901AM "Field Incident Manual." The line reference letters refer to 901AM. Complete information on the National Fire Reporting System is available in NFPA Standard 901. Additional details can be collected on the reverse side or on additional forms.

NATIONAL FIRE PROTECTION ASSOCIATION 470 Atlantic Avenue, Boston, MA 02210

100M-7-73-FP   Printed in U.S.A.

FIGURE 4-13. Basic Field Incident Report. Reprinted with permission from NFPA Form 901F, Copyright 1973, National Fire Association, Boston, MA.

Entries contained in this report are intended for the sole use of the State Fire Marshal. Estimations and evaluations made herein represent "most likely" and "most probable" cause and effect. Any representation as to the validity or accuracy of reported conditions outside the State Fire Marshal's office, is neither intended nor implied.

STATE OF CALIFORNIA
OFFICE OF THE STATE FIRE MARSHAL

FIRE INCIDENT REPORT

INCIDENT NO.

DEL 1   CORR 2

_____ **FIRE DEPARTMENT**

(DEPARTMENTAL USE)

1. OCCUPANT NAME | RELATIONSHIP | ALARM SOURCE | TEL. BOX | PFAS VERBAL | RADIO OTHER
2. ADDRESS | ROOM / APT. NO. | CITY | ZIP | TELEPHONE NO. (CALL BACK)
3. OWNER NAME | ADDRESS | CITY | ZIP | CENSUS/PARCEL NO.
4. MANAGER NAME | ADDRESS | CITY | ZIP | TELEPHONE NO.

A. INFORMATION (PAGE 17)

1. FIRE DEPT. ID | INCIDENT NO. | EXPOSURE NO. | TIME | MONTH | DAY | YEAR | DAY CODE | COUNTY OF FIRE | DIST/ CITY | OUT OF JURISDICTION — CHECK IF YES

B. PROPERTY CLASSIFICATION (PAGE 19)

1. CODE | TYPE OF INCIDENT | CONSTR DATE PRE 72 / POST 71
2. CODE | PROPERTY CLASSIFICATION (COMPLEX)
3. CODE | PROPERTY CLASSIFICATION (INDIVIDUAL)

C. PROPERTY TYPE (PAGE 41)

1. PROPERTY MANAGEMENT | PVT 1 | FED 2 | STATE 3 | COUNTY 4 | CITY 5 | DISTRICT 6 | FOREIGN 7 | OTHER 8
2. CODE STRUCTURE, BUILDING OR VEHICLE - PROPERTY TYPE | BUILDING NO. STORIES
3. STRUCTURE, BUILDING OR VEHICLE - CONSTRUCTION TYPE | EXT. WALL N/C 1 / COMB 2 | INT. WALL N/C 3 / COMB 4 | FLOOR - ROOF N/C 5 / COMB 6 | FIRE RATED YES 7 / NO 8

D. EXTENT OF DAMAGE (PAGE 45)

1. CODE | EXTENT OF DAMAGE – FIRE
2. CODE | EXTENT OF DAMAGE – SMOKE
3. CODE | EXTENT OF DAMAGE – WATER
4. ESTIMATED LOSS - PROPERTY | ESTIMATED LOSS - CONTENTS

E. LOCATION & CAUSE (PAGE 49)

1. CODE | LEVEL OF ORIGIN
2. CODE | SOURCE OF HEAT CAUSING IGNITION
3. CODE | FORM OF HEAT CAUSING IGNITION
4. CODE | ACT OR OMISSION CAUSING IGNITION

F. AREA, MATERIALS & SMOKE SPREAD (PAGE 63)

1. CODE | AREA OF ORIGIN
2. CODE | TYPE OF MATERIAL FIRST IGNITED
3. CODE | FORM OF MATERIAL FIRST IGNITED
4. CODE | MAIN AVENUES SMOKE SPREAD

G. SPREAD OF FIRE (PAGE 77)

1. CODE | MAIN AVENUES FIRE SPREAD
2. CODE | TYPE MATERIAL CAUSING SPREAD
3. CODE | FORM MATERIAL CAUSING SPREAD
4. CODE | ACT OR OMISSION CAUSING SPREAD

H. PROTECTION FACILITIES (PAGE 91)

1. CODE | SPRINKLERS – TYPE
2. CODE | SPRINKLERS – EFFECTIVENESS
3. CODE | STANDPIPES – TYPE
4. CODE | STANDPIPES – EFFECTIVENESS
5. CODE | PORTABLE EXTINGUISHERS – TYPE
6. CODE | PORTABLE EXTINGUISHERS – EFFECTIVENESS

I. PROTECTION FACILITIES (PAGE 97)

1. CODE | PRIVATE BRIGADE – TYPE
2. CODE | PRIVATE BRIGADE – EFFECTIVENESS
3. CODE | SPECIAL HAZARD PROTECTION – TYPE
4. CODE | SPECIAL HAZARD PROTECTION – EFFECTIVENESS
5. CODE | SIGNAL OR WARNING SYSTEM TYPE | CODE | EFFECTIVENESS
6. CODE | SIGNAL WARNING SYSTEM - MEANS OF ACTIVATION
7. CODE | SIGNAL/WARNING SYSTEM - TYPE DETECTORS
8. CODE | WATCHMAN EFFECTIVENESS | CODE | OTHER FACILITIES EFFECTIVENESS

J. MISCELLANEOUS (PAGE 109)

1. FIREFIGHTER: NO. INJURED | NO. OF DEATHS | CIVILIANS: NO. INJURED | NO. OF DEATHS
2. SFM FORM 60-1 SUBMITTED FOR EACH DEATH (CHECK BOX IF YES) ☐

SFM FORM 60-60 (7/73)

FIGURE 4-14. Fire Incident Report. Courtesy of the Office of the State Fire Marshal, State of California.

K. RESCUE OR INHALATOR

| NAME OF PATIENT | | AGE | SEX | TELEPHONE NO. |

ADDRESS | PARENT OR GUARDIAN

CONDITION ON ARRIVAL: GOOD ☐ FAIR ☐ POOR ☐
CONDITION ON DEPARTURE: GOOD ☐ FAIR ☐ POOR ☐
CAUSE OR AILMENT

AID GIVEN: INHALATOR ☐ HRS_____ MIN_____ RESUSCITATOR ☐ HRS_____ MIN_____ OTHER_____

AMBULANCE | PHYSICIAN | TAKEN TO

L. FIREFIGHTING APPARATUS

COMPANY	RESPONSE TIME	SHUT-OFF TIME	OFFICERS	FIREMEN			PUMPERS RESPONDING	PUMPER CREW WORKED	TRUCKS RESPONDING	TRUCK CREWS WORKED
				ON DUTY	OFF DUTY	VOL.				

EQUIPMENT LOST OR DAMAGED | EQUIPMENT NEEDED BUT NOT AVAILABLE

M. HOSE & LADDERS

HOSE LINES	1"	$1\frac{1}{2}$"	$2\frac{1}{2}$"	OVER $2\frac{1}{2}$"	BATTERY	LADDERS
NO. OF LINES						
NO. OF FEET						

N. MISCELLANEOUS INFORMATION

WETTING AGENT _____ GALS. | FOAM _____ GALS. | WATER USED _____ GALS. | WEATHER COND.

MUTUAL AID | POLICE UNITS | FIRE PREVENTION | PHOTOS BY

IF EQUIPMENT INVOLVED IN IGNITION: TYPE | YEAR | MAKE | MODEL

INSURANCE COMPANY NAME OR AGENT | PHONE | AMOUNT OF COVERAGE: CONTENT | PROPERTY

INSURANCE COMPANY ADJUSTER | PHONE | POLICY NO.

ESTIMATED VALUE: CONTENT | PROPERTY | ACTUAL LOSS: CONTENT | PROPERTY

O. VEHICLE

TYPE OF VEHICLE | MAKE | MODEL | YEAR | LICENSE NO. | STATE

REGISTERED OWNER | ADDRESS | TELEPHONE NO.

Entries contained in this report are intended for the sole use of the State Fire Marshal. Estimations and evaluations made herein represent "most likely" and "most probable" cause and effect. Any representation as to the validity or accuracy of reported conditions outside the State Fire Marshal's office, is neither intended nor implied.

SIGNATURES
PREPARED BY: | DATE
APPROVED BY: | DATE

OSP

FIGURE 4-15. Fire Incident Report. Courtesy of the Office of the State Fire Marshal, State of California.

FIRE DEPARTMENT FIELD INCIDENT REPORT

SECTION A

Incident No.	Exp.	Auxiliary Trip (Check)	Date and Time of Incident	Actual Incident Type	F.D. Action
5 0					

Building No. | Street Name (Print) | | Census Tract | Dist. | Out of Juris. (Ck.)

Property Name | Room No. | Complex (If property is part of a complex.) | Fixed Property Use | Property Mgmt.

Owner/Res. Owner/Manager/Patient | Age | Residence Phone | Bus. Phone | Address (to locate following emergency)

Occupant Driver/Employee/Relative | Age | Residence Phone | Bus. Phone | Address (to locate following emergency)

R.P./Witness/Person Discovered Fire | Age | Residence Phone | Bus. Phone | Address (to locate following emergency)

B If the incident was coded as a fire (a one in box 11 above) complete this section. If not leave blank.

5 1

Area of Origin (KAA) | Equipment Involved in Ignition (KBA) | Form of Heat of Ignition (KBB)

Bldgs. with limits / Outside No Limits
201. basements
202. subbasements, etc.
101. 1st floor
102. 2nd floor
103. 5th floor
401. grade to 9'
402. grade to 9' below
403. 10-19'
404. 10-19' below
405. 10'-49' below

Probable Type Material Ignited (KCA) | Form of Material Ignited (KCB) | Act or Omission—Origin of Fire (KDA)

If there was rapid, intense, or unusual fire spread or smoke spread conditions, complete this section. If not leave blank.

Main Avenue of Fire Spread (LA) | Type of Material Ignited Responsible for Spread (LBA) | Form of Material Most Responsible for Fire Spread (LBB) | Act or Omission of Fire Spread (LCA) | Code Violation

F If incident was a structure fire, complete this section. If not leave blank.

5 2

Nearest Usable Pub. Hydrant | Nearest Usable Priv. Hydrant
1=within 500 feet   4 within 500 feet
2=within 1,000 feet 5 within 1,000 feet
3=over 1,000 feet   6 over 1,000 feet
                     7 no usable hydrant

If fire protection equipment or systems were USED or FAILED TO OPERATE, complete this section. If not leave blank. Any line started must be completed.

Sprinklers in Fire Area — Number | No. Open'd | Effect
1. Wet 4. Deluge
2. Dry 5. Comb.
3. Pre-act.
Effect: 1. No help 4. Ext'g'd 2. Helped 5. Failed 3. Contr'ld

Standpipes — Number | No. Used | Effect
1. Wet 4. Dry
2. Auto 5. Comb.
3. Manual
Effect: 1. No Help 4. Ext'g'd 2. Helped 5. Failed 3. Contr'ld

Extinguishers — Number | No. Used | Effect
2. Water 6. Dry chem.
4. Foam 7. Powder
5. CO₂   8. Comb. ABC
Effect: 1. No help 4. Ext'g'd 2. Helped 5. Failed 3. Contr'ld

Fire Brigade — Size | No. Used | Effect
1. Full-time
2. Part-time
Effect: 1. Fire too small 2. Helped 3. Controlled 4. Extinguished 5. Failed to control

Special Systems | Effect
1. CO₂   4. Water spray
2. Dry Chem. 5. Explos. 8. Halon
3. Foam 6. Static 9. Other
Effect: 1. No help 4. Extinguished 2. Helped 5. Failed 3. Contr'ld

Alarm Systems | Effect
1. Local   4. Proprietary
2. Auxiliary 5. Central Sta.
3. Remote Sta. 6. Household
Effect: 1. No alarm 4. Used but failed 2. Delayed 5. Satisf. 3. Satisf.

Watchman | Effect
(Check if Yes, Blank if No)
Effect: 1. No alarm sent 3. Satisfactory 2. Delayed alarm

E If incident was a structure fire, complete this section. If not leave blank.

5 3

Fixed Property Type
1. Bldg., 1-2 occupancy
2. Bldg., multi-occup.
3. Stadium, arena, open patio, etc. trestle, parking garage, etc.
4. Tent or air supported structure
5. Public stockyards, picnic grounds, etc.
6. Private stockyards, corrals, etc.
7. Mine

| Year Const. (EA) | Height | Sq. Ft. of Ground | Type Const. (EE) |

1. Fire resistive (concrete & 2 hr or 3 hr protected steel, usually hospitals or hi-rise)
2. Heavy timber (brick, block, tiltup walls & heavy timber supports, 8x8 col., 6x10 beams)
3. Prot. non-comb (brick, block, tiltup or metal walls & prot. steel supports, prot. metal roof)
4. Unprot. non-comb (same as No. 3 except supports & roof not protected, some service stations)
5. Prot. ordinary (brick, block, tiltup walls & 1 hr prot. wood supports, 1 hr prot. wood roof)
6. Unprot. ordinary (same as No. 5 except supports and roof not protected, most LG shop centers)
7. Prot. wood frame (5/8" sheetrock or 3/4" plaster int., stucco ext. or sheetrock & brick/wood)
8. Unprot. wood frame (wood frame bldg., less than 5/8" sheetrock, most homes, apts., offices, etc.)

Most Recent Inspection
| Month | Day | Year | Violation Noted? (Check if Yes) |
Contributed to Fire? (Check if Yes) ☐ 41
Obstacles to Rescue ☐ 42

Fire/Heat Dmg. Confined to: (TA)
1. First material ignited
2. Part of room of origin
3. Room of origin
4. Fire floor
5. Bldg. of origin
6. Beyond bldg. of origin

Smoke/Water Dmg. Confined to: (TB)
1. First material ignited
2. Part of room of origin
3. Room of origin
4. Fire floor
5. Bldg. of origin
6. Beyond bldg. of origin

Days out of Business

C

Fire Spread Beyond Designed F.P. Limits (TC)
11 Fire loading
12 Fire door open
13 Window open
14 Holes in fire wall
15 Mechanical failure
16 Lack of management
17 Design, const./inst. deficiency
18 Delayed detection
19 Failure to isolate fire
20 Non-rated elements
21 Natural elements
22 Other

D If the incident resulted in any fire loss, complete this section.

5 5

	Estimated Value	Estimated Loss (Fire/Heat/Smoke/Water Dmg.)
Structure or Vehicle		
Contents or Cargo		

Insurance/Agent/Broker/Co.

If mobile property fire, complete this section. If not leave blank.

5 4

| Mobile Property Use (DC) | License | State |
| Make | Year | Color | Model |

FIGURE 4-16. Fire Department Field Incident Report.

FIRE DEPARTMENT FIELD INCIDENT NARRATIVE
(Use paragraph numbering to organize your report. Use additional pages if necessary.)

1. FIRE DEPARTMENT ACTION—Give a complete account of the work of the company(ies) upon FD arrival. Note the sequence of operations and which floor or location the crews (companies) worked. Where appropriate make a sketch showing relative positions of apparatus, hydrants, ladders, lines of attack and which men (companies) manned them, fire building, exposures, etc.
2. UNUSUAL CIRCUMSTANCES—Note the following: condition of doors or windows (locked, open, etc.); suspicious conditions or obstacles; names, addresses, descriptions, etc. of witnesses or suspects. Match names of FD personnel who found these conditions.
3. STORY OF THE ALARM—Briefly tell the story of the alarm prior to FD arrival; note how it occurred, how it progressed, who was involved, how the alarm was reported, and your personal observations on arrival. If the information is from another person, identify him. If the information is your opinion, list the facts influencing or supporting this opinion.
4. ADDITIONAL DETAILS—List all pertinent facts regarding additional alarms; names of Duty Chief or Inspector; Police badge numbers and ambulance or hospital names; facts regarding cause, nature, extent of injured parties; descriptions of extent of damage, etc.
5. The OIC completes this report, using supplementary reports as needed, e.g.: FD 200A, 300, 500, or 500P, which are completed by the individual company officers.

Check Officer in Charge	Narrative Prepared by	Coding Prepared by (Check)
✓ Print Name, Assignment	✓	Signature ✓
B/C		
Captain		
Other		

(Coding Based on 73 NFPA 901)

FIGURE 4-17. Fire Department Field Incident Narrative.

Fire departments respond in many types of incidents. Each incident is a unit of work and should be recorded. Some jurisdictions report that 15 percent of their total incidents are fire-related after investigation. Others have a higher percentage of fire-related calls than this, but all incidents must be reported in order to identify the degree of work involved.

Figures 4-10 and 4-11 show the times of specific alarms, who called them in, and so forth. These are manual records and data processing forms may utilize the same criteria. Figure 4-18 is an example of a UFIRS-type alarm report that enables the department to compute statistics on each fire unit, or to evaluate the various methods people may use to report fires or other emergencies, for example, telephone, fire alarm box, and so forth. This report is usually completed by the dispatcher, who may not be a fire fighter or even in the fire department. Radio enables the dispatcher to enter the necessary time elements on the document as they occur. The computerized report then summarizes the data in the format desired and it also incorporates the data collected by the fire department's OIC.

Company Run Reports

These reports are made out by the officer in charge (OIC) of a company responding to an alarm or incident. The OIC reports only the operations of his or her own company, unless, of course, the particular OIC was in overall command. This report is completed immediately after the incident and should contain all information including drawings and sketches that can be used later to fully explain what was observed, where it was observed, and what was done by the company. These are often invaluable later in analyzing operations, developing better methods of attack, and aiding cooperation between companies. Figure 4-19 is an example of a UFIRS Company Incident Report that would accompany the OIC's Incident Report (Figure 4-16).

Today's specialized rescue and emergency medical service activities require additional data requirements. Figures 4-20 and 4-21 are examples of one set of data elements required for a paramedic system operated by the Palo Alto Fire Department, in cooperation with the Stanford Medical Center Emergency Medicine Department. These forms are used to identify a patient's condition in the field and they record all field treatment and conditions until the patient is released to the doctor at the hospital emergency room who has been communicating via radio with the field paramedics.

The increasing need for more and better emergency medical systems everywhere has generated a new requirement for data. Reports of almost

Incident Number	Supp.
4 0	
1 2 3	7 8 9

Time and Date Call Received
Time	Month	Day	Year
10	13 14		19

LOCATION/ADDRESS   CROSS STREET   DISTRICT   MAP NO.

INCIDENT TYPE: _____

Alarm Source (20)

1. □ Telephone
2. □ Box
3. □ PFAS
4. □ Verbal/in Person
5. □ Radio
9. □ Other

REPORTED BY: _____ PHONE NO. _____

INFORMATION REPORTED _____

_____

BOX NO. TRANSMITTED _____

Box No. Received	First Due Unit	Highest Alarm Level	Check box if more than 10 units are dispatched
			□
21 26	27 31	32	33

(73)
Change □   Delete □
2   3

Time and Date Incident Under Control	Time and Date Last Unit Departed Scene
34 37 38 43	44 47 48 53

Dispatcher(s): _____

4 1	Responding FD Units	Time and Date Unit Dispatched				Time and Date Unit Arrived at Scene				Time and Date Unit in Service				C O D E
1 2	10 14	Time	Month	Day	Year	Time	Month	Day	Year	Time	Month	Day	Year	
Alarm	10 14	15 18	19		24	25 28	29		34	35 38	39		44	45

Note: A separate card must be punched for each row of data. Positions 1-2 contain the card code, and positions 3-9 contain the incident and supplemental numbers.

4 1	M E D 1	MEDICAL SERVICE
1 2	10 14	

If not dispatched—
unit disposition (67)
1. □ Out of service
2. □ On other call

	Time and Date Unit Dispatched				Time and Date Unit Arrived at Scene				Time and Date Unit Arrived at Hospital				C O D E
	Time	Month	Day	Year	Time	Month	Day	Year	Time	Month	Day	Year	
	15 18	19		24	25 28	29		34	35 38	39		44	45

Caller reports (68)
1. □ Possible heart attack
2. □ Drowning
3. □ Unconscious
4. □ Breathing problem
5. □ Seizure
6. □ Major injury
7. □ Possible dead body
8. □ Minor injury
9. □ Overdose—poison
0. □ Other _____

Dispatched from (69)
1. □ Station
2. □ Structure call
3. □ Paramedic call
4. □ Hospital
5. □ Field
6. □ Other

Time and Date Unit Left Scene				Time and Date Unit in Service			
Time	Month	Day	Year	Time	Month	Day	Year
47 50	51		56	57 60	61		66

Proceeded to (70)
1. □ Hospital
2. □ Station
3. □ Another call

ALARM REPORT

FD-100
7/75

FIGURE 4-18. Alarm Report.

COMPANY INCIDENT REPORT

FD-300

COMPLETE FOR ALL INCIDENTS

6 | 0 | | | | | | 7 | 8 9 10 | | | 14 | | |
1 2 3   7 8 9 10   14

INCIDENT NO.   SUPP   COMPANY NO.

(73)
CHANGE 2 ☐   DELETE 3 ☐

A

| | 15 16 |
OFFICERS (ON DUTY)

| | 17 19 |
FIREMEN (ON DUTY)

| | 20 21 |
OFFICERS (SPECIAL CALL)

| | 22 24 |
FIREMEN (SPECIAL CALL)

PUMPERS RESPONDED — 25 26
PUMPER CREWS WORKED — 27 28
PUMPS USED — 29 30

NO. MEDICAL UNITS RESPONDED — 31 32
CREWS WORKED AS MEDICS — 33 34
CREWS WORKED AS FIREFIGHTERS — 35 36

AERIAL LDRS/PLATFMS RESPONDED — 37 38
AL/P CREWS WORKED — 39 40
AL/P USED — 41 42

OTHER APPARATUS RESPONDED — 43 44
OTHER APPARATUS CREWS WORKED — 45 46
OTHER APPARATUS USED — 47 48

COMPLETE IF EQUIPMENT OR CHEMICALS WERE USED

6 | 1
1   2   FIRE STREAM REPORT

B

¾″ OR 1″ HAND LINES — 15 16
1½″ OR 2″ HAND LINES — 17 18
2½″ OR 3″ HAND LINES — 19 20
MASTER STREAM APPLIANCES — 21 22
OTHER: — 23 24

SOLID STREAMS — 25 26
SPRAY/COMBINATION STREAMS — 27 28

MAX. FLOW RATE (GPM) — 29 32
DURATION OF FLOW (MIN) — 33 36
TOTAL WATER (100 GAL.) — 37 40

EQUIPMENT USED REPORT

HOSE (HUNDREDS OF FEET) — 41 42
LADDERS UP TO 30′ — 43 44
LADDERS OVER 30′ — 45 46
HAND EXTINGUISHERS — 47 48
MASKS — 49 50
SALVAGE COVERS — 51 52

SMOKE EJECTORS — 53 54
ELECTRIC GENERATORS — 55 56
RESUSCITATOR — 57 58
PORTABLE PUMPS — 59 60
FLOODLIGHTS — 61 62
POWER SAWS — 63 64
HURST TOOL — 65 66

6 | 2
1   2   CHEMICALS REPORT

CARBON DIOXIDE (LBS.) — 15 18
DRY CHEMICALS (LBS.) — 19 22
FOAM/SURFACTANT (GAL.) — 23 26

HALOGENATED AGENT (LBS.) — 27 30
OTHER: — 31 34

NOTES: POSITIONS 3–9 OF EACH CARD MUST CONTAIN THE INCIDENT AND SUPPLEMENTAL NUMBERS.
POSITIONS 10–14 OF EACH CARD MUST CONTAIN A COMPANY NUMBER.

C

REMARKS:

COMPANY OFFICER

6/75 50M FP 50M

2-71-50M-FP

FIGURE 4-19. Company Incident Report.

CASUALTY REPORT — 1

PAGE   OF

(50) FIRE ☐
(51) NON-FIRE ☐

(73)

Change 2☐   Delete 3☐

(Check One)

Month	Day	Year	Time
50		55	

Incident No.	Supp.	I.D.	Name		Age
8 | 0				33 34 36	
1 2 3	7	8 9	10 11 12		

Sex Male (M)   Female (F)   Injury (I) or Death (D)   Civilian (C) or Firefighter (F)   Fixed Property Use   DB   Mobile Property Use   DC

37   38   39   40   42   43 44

(45) Condition before Casualty

☐ 1 Asleep
☐ 2 Bedridden or other physical handicap
☐ 3 Impaired by drugs or alcohol
☐ 4 Under restraint
☐ 5 Too young to act
☐ 6 Too old to act; senile
☐ 7 Mentally handicapped
☐ 8 Awake and unimpaired
☐ 9 Other (specify)
☐ 0 Undetermined or not applicable

(46) Action Causing Casualty

☐ 1 Caught in, under or between or trapped by
☐ 2 Exposed to: heat, chemicals, radiation, smoke, etc.
☐ 3 Fell over, on, or tripped on
☐ 4 Stepped on or into
☐ 5 Overexertion
☐ 6 Rubbed by or contact with
☐ 7 Struck by
☐ 8 Not applicable
☐ 9 Other (specify)
☐ 0 Undetermined

(47) Nature of Casualty

☐ 1 Burns and asphyxia/smoke
☐ 2 Burns only
☐ 3 Asphyxia/smoke only
☐ 4 Wound, cut, bleeding
☐ 5 Dislocation, fracture
☐ 6 Complaint of pain
☐ 7 Shock
☐ 8 Strain, sprain
☐ 9 Other (specify)
☐ 0 Undetermined

(48) Part of Body Injured

☐ 1 Head, neck, includes respiratory system
☐ 2 Body, trunk, back
☐ 3 Arm
☐ 4 Leg
☐ 5 Hand
☐ 6 Foot
☐ 7 Internal except respiratory system
☐ 8 Multiple parts
☐ 9 Other (specify)

(49) Disposition of Casualty

☐ 1 Refused help
☐ 2 First aid at scene and released
☐ 3 Taken to hospital—by fire department vehicle
☐ 4 Taken to hospital—by non-fire department vehicle
☐ 5 Taken to other than hospital
☐ 6 Died
☐ 9 Other (specify)
☐ 0 Undetermined

8 | 4		Oxygen Kit	Airway Kit
1 2			

Oxygen Kit

(20) ___ Oxygen supply tubing
(21) ___ Nasal cannula, child
(22) ___ Nasal cannula, adult
(23) ___ Pediatric oxygen mask
(24) ___ Adult oxygen mask

Airway Kit

(25) ___ Laerdal RFB 11 Resuscitator
(26) ___ Adult size mask
(27) ___ Child size mask
(28) ___ Esophageal obdurator with mask
(29) ___ Lubricant packet
(30) ___ Syringe, 2 oz. with luertip
(31) ___ Nasal pharyngeal airway, adult

(32) ___ Nasal pharyngeal airway, child
(33) ___ Magill forceps, adult
(34) ___ Tongue forceps
(35) ___ Mabrick clamp
(36) ___ Yankauer suction tube
(37) ___ Bulb syringe
(38) ___ Nasal gastric tube

Oral Pharyngeal Airway 39 40   Suction Catheter 41 42

8 | 5	
1 2	

Symptoms First Noted

Time	Month	Day	Year
12	15 16		21

(22) Attended by

☐ 1 Ambulance
☐ 2 Civilian
☐ 3 Nurse
☐ 4 Physician
☐ 5 Parent/family

Care Provided
Before Arrival of Fire Department

(23) ☐ Airway
(24) ☐ Bandaging
(25) ☐ Breathing
(26) ☐ Control Bleeding
(27) ☐ CPR
(28) ☐ Defibrillation
(29) ☐ Drew Blood
(30) ☐ Drug
(31) ☐ Irrigation
(32) ☐ IV
(33) ☐ Oxygen
(34) ☐ Psychiatric Care
(35) ☐ Public Relations Only
(36) ☐ Spinal Immobilization
(37) ☐ Splinting
(38) ☐ Vitals
(39) ☐ Other

Action by Paramedic

☐ (40)
☐ (41)
☐ (42)
☐ (43)
☐ (44)
☐ (45)
☐ (46)
☐ (47)
☐ (48)
☐ (49)
☐ (50)
☐ (51)
☐ (52)
☐ (53)
☐ (54)
☐ (55)
☐ (56)

I.V. Equipment   (49) ___ Sodium chloride 1,000cc

(43) ___ Metri-set
(44) ___ Arm board, long
(45) ___ Arm board, short
(46) ___ Blood set Y-TYPE   I.V. Needles   50 51
(47) ___ D5W 500cc
(48) ___ LR 1,000cc   Small Vein Infusion Set   52 53

Miscellaneous

(54) ___ EKG electrodes
(55) ___ Hot pack
(56) ___ Cold pack   Syringe   57 58

PATIENT CONDITION UPON ARRIVAL

(57) Mental State	(58) Skin Temperature	(59) Skin Color	(60) Pupils		Respiratory	(67) Speech		Pulse

(57) Mental State

☐ 1 Conscious
☐ 2 Unconscious
☐ 3 Stuporous

(58) Skin Temperature

☐ 1 Normal
☐ 2 Hot
☐ 3 Cold

(59) Skin Color

☐ 1 Normal
☐ 2 Cyanotic
☐ 3 Pale, ashen
☐ 4 Flushed

(60) Pupils

☐ 1 Responsive
☐ 2 Sluggish
☐ 3 Unequal
☐ 4 Pinpoint
☐ 5 Dilated, fixed
☐ 6 Unequal - Responsive
☐ 7 Unequal - Sluggish

Respiratory

(61) ☐ Below 10
(62) ☐ Above 20
(63) ☐ Labored
(64) ☐ Shallow
(65) ☐ Moist
(66) ☐ Wheezing

(67) Speech

☐ 1 Coherent
☐ 2 Incoherent
☐ 3 Hysterical
☐ 4 Silent

Pulse

(68) ☐ Below 60
(69) ☐ Above 100
(70) ☐ Regular
(71) ☐ Irregular

Time	BP	Pulse	Resp.	Rhythm	Defib.

(72) BP (Systolic)

☐ 1 Below 90
☐ 2 Above 140

(74) Skin Moisture

☐ 1 Normal
☐ 2 Dry
☐ 3 Moist

(75) Coroner's Case

☐ 1 Decapitated
☐ 2 Rigor mortis
☐ 3 No vital signs

Prepared by: _____

FD-500P
7/75

FIGURE 4-20. Casualty Report—1.

CASUALTY REPORT — 2

8	6	Incident No.	Supp.	ID No.	Casualty Name
1	2	3 ... 7	8	9 10 11	

(73)

Change 2 ☐   Delete 3 ☐

(12) ☐ Hospital Communication   1 = Phone   2 = Radio

Cardio
(18) ☐ Chest pain
(19) ☐ Full arrest
(20) ☐ Arrhythmia
(21) ☐ Other (Specify)
_____

CNS
(22) ☐ Seizure
(23) ☐ CVA
(24) ☐ ETOH
(25) ☐ Fainting
(26) ☐ Overdose
(27) ☐ Unconscious
(28) ☐ Diabetic
(29) ☐ Dizzy
(30) ☐ Other (Specify)

Trauma
(31) ☐ Multi trauma
(32) ☐ Stab wound
(33) ☐ Gunshot
(34) ☐ Laceration
(35) ☐ Burns
(36) ☐ Fracture
(37) ☐ Other (Specify)
_____

Abdomen
(38) ☐ Ingestion
(39) ☐ GI bleeding
(40) ☐ Nausea
(41) ☐ Pain
(42) ☐ Other (Specify)
_____

Respiratory
(43) ☐ SOB
(44) ☐ Airway Obst.
(45) ☐ Apnea
(46) ☐ Drowning
(47) ☐ Other (Specify)

Miscellaneous
(48) ☐ Obstetrics
(49) ☐ DOA
(50) ☐ Other (Specify)

Time		Code	Dose	Route	Time		Code	Dose	Route

8	7	Incident No.	Supp.	ID No.
1	2	3 ... 7	8 9	10 11

Drug Codes

01 - Bicarb	04 - 50% Glucose	07 - Atropine	10 - Aminophyiline	13 - Lasix
02 - Epinephrine	05 - Ipecac	08 - Isuprel	11 - Benadryl	14 - Phenobarbital
03 - Narcan	06 - Lidocaine	09 - Morphine	12 - Pitocin	15 - Valium

Medical History
(15) ☐ Cardiac
(16) ☐ Diabetic
(17) ☐ Hypertension
(18) ☐ Allergic
(19) ☐ On medication
(20) ☐ Seizure
(21) ☐ COPD
(22) ☐ Recent surgery
(23) ☐ Asthma
(24) ☐ Other (Specify)

Disposition
(25) ☐ Transported to other hospital   (_____) (26)
(27) ☐ Transported to Stanford, Paramedic
(28) ☐ Transported to Stanford, Private Ambulance
(29) ☐ Transported to Stanford, Non-ambulance
(30) ☐ Paramedic accompanied patient to hospital
(31) ☐ Transported home
(32) ☐ DOA at hospital
(33) ☐ Patient released
(34) ☐ MD office or clinic
(35) ☐ Patient refused treatment
(36) ☐ Other

1 - El Camino
2 - Kaiser
3 - V.A. Menlo Park
4 - V.A. Palo Alto
5 - Sequoia
6 - Coroner

Condition on Arrival at Hospital
(38) ☐ Conscious
(39) ☐ Unconscious
(40) ☐ Shock
(41) ☐ CPR
(42) ☐ Dead

Name of Transporting Company to Hospital
(37)   1 - Palo Alto   3 - Fields   5 - Other
        2 - V.A.   4 - Mercy

Paramedics Names	Code
_____	
_____	

Patient's Home Address: _____

_____

FD-501P
7/75

FIGURE 4-21. Casualty Report—2.

every size and description are in existence now, and we will soon see a Uniform Medical Report developed, similar to the UFIRS report generated by the NFPA.

Patient records for all emergency activities may be privileged under the law. Many areas have laws that extend such privacy to include physician/patient confidentiality. Therefore, where applicable, records may not be released to interested parties as public records, but may only be released by the physicians responsible or the patients themselves.

Other Reports and Records

There are many other forms and records kept by almost every fire department of any size. A good many are related to the department's administration. In this group we find:

1. Budget and finance records, showing balances and expenditures from all accounts in the approved budget, and inventories of supplies, clothing, equipment, apparatus, and so forth, for all units. (See the Finance section in this chapter.)
2. Personnel records, showing payroll, individual service records, retirement and pension records, reports of all disciplinary hearings and actions. (See the Personnel section in this chapter.)

REVIEW QUESTIONS

1. Why is fire department organization patterned after the military?
2. What is meant by "chain of command" and why is it important in fire department organization?
3. What factors are considered in determining the number of members assigned to a company?
4. Describe an engine company organization, giving titles and positions.
5. Name five requirements for a driver of fire apparatus.
6. Why should a member of an engine company understand the duties of a ladder company member?
7. Describe the organization chart of:
    (a) a one-company volunteer fire department;
    (b) a large paid fire department that includes battalions and districts.
8. What are some of the duties of a fire chief that can be considered management?
9. How is management handled in a large department?

10. Why are public relations important to a fire department and how can they be applied?

11. Name the reasons for fire investigations.

12. How can thorough investigations reduce future arson fires?

13. Name seven reasons why planning is important to a fire department.

14. What is group planning and what are its advantages?

15. Personnel departments can improve recruitment standards by assuring that reasonably qualified candidates are referred to the written examinations administered by whom?

16. What are the functions of a training department?

17. Name some of the duties of a fire prevention bureau and some of the difficulties.

18. How can company inspections help and what are the limitations?

19. What is the purpose of a communications bureau?

20. Since the water system is usually under some other governmental unit's control, why should a fire department have a water officer or water bureau?

21. Name seven duties of the service division of a fire department.

22. What are the advantages of the uniform reporting and coding of records of fire incidents?

5 Fire Department Equipment

As we saw in Chapter 4, modern fire department companies center around their major piece of apparatus. Except for trained fire fighters, apparatus is the principal asset of any fire department, regardless of its size. If apparatus is outmoded, in poor condition, or worn out, even the best trained fire fighters and officers will be severely handicapped in doing their most important job: fighting fire.

In short, modern, well-functioning apparatus is the key to successful fire operations. For this reason, although there are entire courses on apparatus and equipment, we will take the time to discuss the subject here in some detail.

Pumpers

Water is our principal weapon against fire involving ordinary combustible materials. It is also the only practical extinguishing agent to combat large-scale fires. The ability of a fire department to carry water from an available source to a fire, and once there to deliver it in such quantities that it will extinguish the fire, depends on three things: the available water supply, the right kinds of hose, and the pumper used to supply water under proper pressures.

Water supply systems will be discussed later in this book, so we will limit our discussion here to pumpers and hose.

Early in the development of organized fire fighting, pumpers were simple devices (Figure 5-1) with two or four persons working a double-handled lever to supply the hose with water under pressure.

The hose itself was formed by stitching leather into a tube and riveting short sections together. When wet, such leather hose was unbelievably

FIGURE 5-1. Early hand pumper. Courtesy of the San Francisco Fire Department. Photo by Chet Born

heavy. It leaked badly, was stiff, hard to couple, and in cold or freezing weather it took on the characteristics of an iron bar. The first major advance in hose came in 1871 when the Cincinnati Fire Department got its first rubber-and-fabric hose. Alternate layers of heavy cotton duck (a kind of canvas) and thin rubber surrounded a central rubber tube, which was protected on the outside by a final rubber coating. This type of hose became the standard until hose with a seamless, circular-woven jacket came into use. Such hose was lighter, waterproof, handled better, and could be coupled quickly, a big advantage in fire fighting.

As buildings started to get taller and bigger, the primitive hand pumper was no longer able to supply enough pressure to raise streams to the point where they were effective in fighting fire, nor could human muscle sustain the demands for water for more than a limited time. In answer to the need for better pumpers, makers of equipment developed the steam pumper (Figure 5-2).

Carried on a horse-drawn vehicle and trailing smoke from its polished firebox chimney as it answered a call, the new apparatus gave the fire department a glamour that it never before had. At the same time, steam-driven pumps were much more efficient and far more powerful. In fact, the steam-driven piston pumps, without pressure regulators and other

FIGURE 5-2. Early steam pumper. Courtesy of the Smithsonian Institution.

controls commonly in use today, proved so powerful that the old leather hose frequently burst or leaked badly. However, hose was improved to meet the demands of increased pressure, culminating in the type of hose described above.

Pumps

Following World War I and the widespread change from horse-drawn vehicles to gasoline-powered equipment, pumpers were built around a rotary pump driven by a takeoff from the pumper's engine. And, as vehicles improved, particularly with the advent of the diesel engine, pumps continued to improve, culminating in the centrifugal pumps that are found in 90 percent of today's pumpers. The greatest factor favoring the shift to centrifugal pumps is that this type of pump can make full use of the pressure supplied from hydrants; it also supplies a steady stream free from pulsation, costs less, and weighs less than its predecessors.

As Figure 5-3 shows, centrifugal pumps depend on "impellers," mounted in a housing with one or more suction inlets, two or more

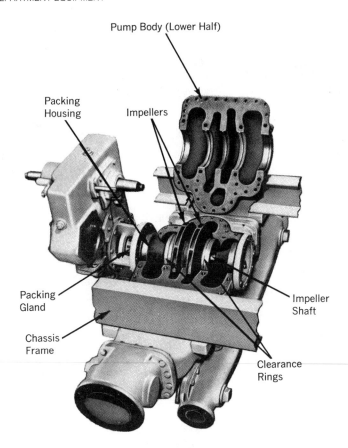

Pump Body (Lower Half)

Packing
Housing   Impellers

Packing
Gland

Chassis
Frame

Impeller
Shaft

Clearance
Rings

FIGURE 5-3. Underframe view of Hale two-stage centrifugal pump
with priming pump removed. Courtesy of the Hale Fire Pump Company.

64 mm (2½ inch) discharge outlets, and a separate transmission system.
Through the gears in this transmission, engine speed can be raised to
the level needed to drive the impeller(s) properly.

When a centrifugal pump is turned on, water from the supply hose
goes to the center of the impeller, called its *eye.* The whirling vanes
pick up this water and whip it outward, increasing its pressure and
velocity. As water leaves the impeller, it creates a partial vacuum at
the eye. Atmospheric (or other) pressure forces water through the suction
hose to replace what has been lost. Water leaving the impeller passes
into a spiral chamber, a *volute,* where it finds a gradually expanding
channel that equalizes the speed of flow and converts the energy of
its high velocity into pressure energy.

Although we have considered a single impeller pump, there are a number of other designs that use one or more impellers to create higher pressures and greater gallonage per minute. These multiple-impeller pumps include the parallel series type, two-stage type, three-stage type, four-stage type, two-stage front-mount type, and the duplex multistage type.

Centrifugal pumps have one drawback: all air must be exhausted from the pump and suction lines before it will work. This means that when a pumper must get water from an unpressurized source, such as a pond, lake, or river, through its suction hose (a process called *operating at draft*), it needs an auxiliary priming device. And if the primer fails, the pump is useless.

There are seven major sizes of centrifugal fire pumps in wide fire service use, with many more options available to meet specific needs. They are:

lpm (liters per min.)	gpm (gallons per min.)	lpm (liters per min.)	gpm (gallons per min.)
1890	500	4731	1250
2840	750	5678	1500
3785	1000	6624	1750
		7570	2000

Note: The conversion factor used is 3.785 liter = 1.000 gallon.

Pumps must pass various acceptance tests that are specified in the NFPA Automotive Fire Apparatus Standard. Generally, these tests include a hydrostatic test by the manufacturer, a 2-hour (run-in) test prior to delivery, and a 3-hour certification test, performed by a recognized testing agency. The latter test establishes the delivery capability of the pumper, at draft, with the driving engine powering the pump. Future service tests are compared with this test to determine efficiency of the pump, its engine or parts. Fire departments are required to test their pumps at least annually, and these service tests are recorded on forms and compared with certification tests.*

The 1892 lpm (500 gpm) pump was one of the most common pumps in earlier years. However, with improved driving engines and improved water supplies available, it is not as effective today. The most common

---

*See Appendix D for samples of test forms in use by various fire departments. A complete study of tests can be found in *Fire Company Apparatus and Procedures* by Lawrence W. Erven, 2nd ed.; published by Glencoe Press, Beverly Hills, California, 1974.

pump capacities in use today range from the 3785 lpm (1000 gpm) size to the 5767 lpm (1500 gpm) size.

There are several models of centrifugal fire pumps with capacities up to 17 030 lpm (4500 gpm). These, of course, are special purpose pumps, and show great promise for future fire service applications. These large pumps may be "built-in" protection devices in new building requirements in the not-too-distant future, as they would be very costly for a local fire department to acquire and maintain. Water systems are also a limiting factor, and these special applications may have stored water as part of the building fire protection systems.

In general, any centrifugal pump that is rated at over 2840 lpm (750 gpm) will be mounted in the center of the vehicle that carries it. While most are powered by a drive that takes off from behind the road transmission, at least two are so designed that the drive can operate when the vehicle is moving, an advantage in fighting brush and forest fires.

Skid-mounted fire pumps are often found on rural fire fighting trucks. These units have a separate power supply engine from that used by the vehicle, which permits pumping in motion and without the limitations usually found with power takeoff connections.

These pumps can be of various capacities and permit the local jurisdiction to use the truck chassis available locally, keeping their overall costs down. Maintenance costs, too, are usually lower, as all piping, connections, valves, and so forth are visible and easily maintained.

Auxiliary pumps are also commonly found on pumper equipment, but they are operated and driven independently, usually by a power takeoff from the side of the road transmission. They are classed as auxiliary because they have less capacity than the minimum standard capacity of 1890 lpm at 1030 kPa (500 gpm at 150 psi). Auxiliary pumps fall into two subclasses: booster pumps, with capacities up to 946 lpm at 827 kPa (250 gpm at 120 psi), and high-pressure pumps that develop 4130 kPa to 5850 kPa (600 to 850 psi) or more with a maximum volume of around 379 lpm (100 gpm).

Booster pumps are self-priming, rotary types. They either connect directly to a water tank on the pumper and feed directly to the booster line or they have two separate 2½-inch suction inlets, one on each side of the pumper, plus the link to the water tank. A relief valve protects the hose line against an excessive pressure rise and against damage to the pump.

High-pressure pumps, used for water fog production, can be centrifugal, as we have seen, but many are the three-cylinder, single-acting piston type. Piston pumps have positive displacement and are self-priming, driven by a power takeoff train mounted on the side of the road transmission. A preset relief valve protects the system against excessive pressure and these pumps can be operated at draft as well as from a water tank on the pumper.

While pumper equipment varies from department to department, every pumper carries fire hose that takes its water from the pump and hose to supply water to the pump, either when operating at draft or off a hydrant. A standard pumper body must have hose compartments that can carry at least 305 m (1,000 feet) of double-jacketed 2½-inch hose and space for at least 122 m (400 feet) of 1½-inch hose for use inside buildings. It may also carry 61.0 m (200 feet) of rubber covered one-inch hose on a reel. In practice, many pumpers carry 610 m (2,000 feet) or more of 2½-inch hose or even larger sizes. Two or more sections of smooth bore suction hose, a minimum of 6.1 m (20 feet) usually decorate the outside of the pumper. Figure 5-4 shows a modern Diesel pumper.

In addition, pumpers that meet NFPA standards* carry at least one 4.3-m (14-foot) ladder with folding hooks (roof ladder) and one 7.3-m (24-foot) extension ladder. (For ladder details, see next section.) The NFPA also list a variety of basic and required items of equipment to

FIGURE 5-4. Modern diesel pumper.

*NFPA 1901, *Standard for Automotive Fire Apparatus,* 1975 edition.

be carried by fire pumpers, and several desirable items as well. Pumpers serving rural or suburban areas should have extra suction hose, longer extension ladders, and several specialized items used in combating grass and brush fires common in rural areas.

One widely used arrangement deserves special mention: the triple combination pumper. This apparatus carries a pump mounted on the vehicle chassis, 365.8 m to 609.6 m (1200 to 2000 feet) of 2½-inch jacketed hose, a large water tank, up to 15 m (50 feet) of ground ladders, miscellaneous small tools, and a wide variety of nozzles.

One advantage of a triple combination pumper over those having no booster pump or water supply is that the officer in charge has more flexibility in decisions regarding attacking a fire and holding or extinguishing it immediately, or laying hose to or from a water supply first. This can be particularly valuable for incipient fires and in rural areas where water supplies are scarce. (See Figure 5-5.)

FIGURE 5-5. Triple combination pumper at fire scene, ladders removed. Courtesy of the Boulder, Colorado, fire department.

Improved pump design and greater pump power are available today at lower costs. These technological advances were only dreamed about in years past. However, they have introduced some severe problems into the fire service.

More crew members are required to operate these new pumpers, and employee costs continue to soar each year. In many cases these costs have reduced the crew level of pumpers to where the capacity of the pumps cannot be fully utilized, instead of the increased pumping capacity resulting in more water being applied in an improved fashion at fires.

The *Grading Schedule for Municipal Fire Protection,* issued by the Insurance Services Office (ISO) has reduced the older requirement of seven members per pumper to six members, and has provided increased credits for crew sizes. But costs are continuing to influence staffing levels adversely.

Every effort should be made to educate the various publics about the effects of reduced crew levels, and possible solutions. The future may see municipalities requiring buildings with built-in fire protection systems, shifting the economic burden to those who require the maximum fire protection for themselves and/or for their communities' safety.

Grass or Brush Fire Trucks

Ordinary heavy-duty pumper apparatus may not be adapted to brush fire work and sending it out on such a run may be needlessly costly. To meet the need for equipment to fight fires in grass, brush, rubbish, and vacant lots, many municipalities and small communities have added special trucks and equipment. Sometimes called a community forest fire truck, the brush fire truck carries a 757-liter to 1890-liter (200- to 500-gallon) water tank on a small commercial chassis (in rough country or hilly terrain, four-wheel drive is advantageous), a booster pump with 757 lpm (200 gpm) capacity, and a large amount of small diameter hose. (See Figure 5-6.) These trucks, or an auxiliary pickup truck, also carry a load of forestry hose and a portable pump to allow use of such water sources as local ponds, small streams, and swimming pools. Many departments also mount a 1890 to 2840 lpm (500 to 750 gpm) two-stage centrifugal pump on the front of their brush and grass fire apparatus. Particularly when equipped this way, the brush fire truck is often used as a booster truck for small stream work on structural fires.

Supplementary equipment includes knapsack-type pump tank extinguishers, hay forks, fire brooms, long-handled spades, metal rakes, and other items suited to fighting forest fires.

Since brush and grass fire vehicles, particularly if used in rural areas, must largely depend on the water they carry, the temptation is great

FIGURE 5-6. Surplus 6 × 6, converted to community forest-fire truck.
Courtesy of the Palo Alto Fire Department.

to carry a tank that is much too large for the vehicle's chassis. When this is done, the resulting overloading of the vehicle can cause loss of stability, favoring overturning and rapid breakdown of the apparatus.

In rural service it is better to have a tanker truck, carrying 5678 liters (1500 gallons) of water and equipped with connecting hose and a pump to fill its own tank or the brush fire truck's tank, accompany the fire apparatus. Tankers can frequently be supplied by local street cleaning departments or contractors.

Airport Crash Trucks

Fighting a fire at an airport calls for highly specialized pumpers. Airplanes, whether jets or propeller driven, carry large amounts of flammable liquids that burn with a fierce, hot flame; parts of the aircraft itself are made of combustible materials and metals. Water alone is of little use against such fires; in fact, it may be dangerous. The proper response to a burning plane is a fast attack with large amounts of foam and chemical agents. (See Figure 5-7.)

To mount this type of attack, airport crash trucks carry their own supply of extinguishing agents and are able to move rapidly over soft

FIGURE 5-7. Airport crash truck. Courtesy of the American LaFrance Company.

ground, dirt surfaces, snow, ice, and the paved runways. Turret nozzles and ground-sweep nozzles deliver foam at high discharge rates; supplementary equipment can deliver carbon dioxide or dry chemical extinguisher. Tanks and supplies of extinguishing agents are big enough to allow about five minutes of continuous operation before backup apparatus can arrive. Airport crash trucks come in two sizes: a major engine, weighing four tons or more, and light rescue vehicles that carry rescue tools and a smaller amount of extinguishing agents for a quick attack on the fire before the heavier equipment can get to the site.

The NFPA lists at least 20 items of crash truck equipment directly related to the problems that are likely to face those fighting an airport fire, plus the same first aid equipment commonly carried for use at any fire.

Hose Carriers

At many fires, more hose is needed than a single pumper can carry. A 2840 liter (750 gallon) pumper can easily handle three lines, each 213 m (700 feet) long, or a total of 183 m (600 feet) more than the standard load carried by many departments on a single engine. Generally, additional hose is made available by assigning two combination pumpers to the same call. Each carries a full load of hose. To make full use of pumping capacities, good practice calls for a reserve hose carrier, with at least 304.8 m (1000 feet) of 3-inch hose to supplement the hose loads of every five to eight pumpers in service in a given department.

Hose

Since no pumper can be better than the hose it carries, let us briefly examine some aspects of hoses, hose fittings, and hose care. Most modern fire hose is made from synthetic rubber, giving it far greater resistance than natural rubber to deterioration caused by ozone (a form of oxygen), weather, oil, and fuels. The exterior jacket of circular woven, seamless fabric can be one of many different fibers; some synthetic, like nylon, and others natural, like cotton. Each kind of fabric has advantages and disadvantages, but in general, all kinds benefit from being dried after use. Fresh from the factory, completely pressure-tested hose can be expected to stand up to any normal demand met in the course of fire fighting. With time and use, however, hose develops defects that must be corrected if the pumper crew is to be able to use it. A burst or broken line can be a disaster by giving a fire enough time to make real headway; it can cost lives when fire fighters are left without water to protect them or contain the flames.

Common causes of damage to hose are chafing, running over it when it is empty or under only low pressure, freezing, and cuts from glass, nails, or other sharp objects. Heat attacks hose in two ways: embers will injure the fabric jacket; exposure to high temperatures causes the inner rubber lining to harden and crack. Chemical attack on the lining comes from gasoline, oils, grease, and paint; acids destroy the fiber jacket.

Couplings are usually the expansion ring type, made of brass composition. There are at least six different kinds of fittings used to tighten couplings. The important thing to watch for in couplings is injuring the threads by forcing, or overlooking dirt or abrasive matter, or by striking the coupling on a hard surface. This forces the metal out of round as well, making a watertight fit impossible.

Most hose problems can be prevented by a program of care that includes thorough inspection after any use. Damage details should be entered in a hose record book (see Appendix D), followed by washing, draining, and drying of the hose. Another important part of hose care is changing the arrangement of hose stored on a pumper so that the position of any bends or folds is shifted from place to place. This will prevent fatigue of the linings and cracking when straightened and put under pressure.

Ladder Trucks

Just as pumpers and pumps have changed greatly since their beginnings, so too have ladders. It might even be argued that no other area of fire science has seen such great changes in equipment in less than half a century. The reasons for these changes are the same as those behind the improvements in pumping apparatus: the change in the height and characteristics of most buildings and the growth of our cities. Since the rescue of people from a burning building is the primary responsibility of a ladder company, we can easily see that the higher the buildings, the longer the ladders must be to reach their upper stories. But the longer the ladder, the heavier it becomes and the more difficult it becomes to move or handle. With this in mind, let us consider the different types of ladders in use today.

Ground Ladders

Ground ladders, in sizes from 3.1 to 17 m (10 to 55 feet) long, come in a variety of forms: wall, extension, hook (roof), and attic (folding). These ladders are used for rescue, to stretch lines into a fire building, to provide ventilation by giving access to ports, scuttles, windows, roofs, or other places that are hard to reach. The shorter lengths find their best use in such emergencies as train wrecks, accidents on utility poles, building collapse, and in rescuing people in danger of drowning.

Originally, ladders were built of hard wood, with solid beams. Even in short lengths, these ladders were heavy and hard to handle. When steel reinforcing rods were added for strength and metal "shoes" put at the top and bottom to protect the wood against damage and to make placement more secure, ladders grew even heavier. Eventually, to overcome the limit sheer weight imposed on handling and carrying, ladder builders brought out the trussed beam. At first, these were of wood; then they were made of lightweight metal alloys. Figure 5-8 illustrates the difference between a solid beam and a trussed beam ladder.

Solid beam ground ladders can still be found in service, even some of wood, but for ladders more than 6.1 m (20 feet) long trussed beam construction is preferred. Not only are trussed beam ladders easier to handle but they also put less weight on the truck that must carry them. Since a ladder truck that meets NFPA standards will carry at least seven ladders of different types and sizes adding up to 49.7 m (163 feet), any weight that can be saved is all to the good.

Straight ladders, regardless of what they are used for, require far more people to handle them than one might think. For example, it may take as many as six people to carry one and raise it into position. When in position, a straight ladder must be securely placed, have someone bracing the foot, or be lashed or dogged in place.

Extension ladders consist of a lower and upper "fly" that can be extended to the desired height either by a halyard (a rope and pully device) or by hand, or by pike pole. These ladders resemble those used

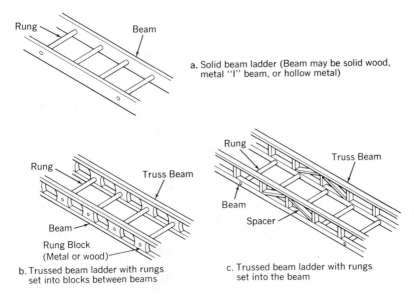

a. Solid beam ladder (Beam may be solid wood, metal "I" beam, or hollow metal)

b. Trussed beam ladder with rungs set into blocks between beams

c. Trussed beam ladder with rungs set into the beam

FIGURE 5-8. Solid and trussed beam ladders.

by painters and carpenters, but are much stronger because of the higher fire service standards. They may be of wood or metal, with solid or trussed beams. Safety hooks or pawls generally engage the rungs of the lower section, once the fly is extended as far as needed, to prevent slipping or failure. Figure 5-9 shows the 11-m (35-foot) length, the workhorse of the average fire department, that can be used at any height from 4.6 to 10 m (15 to 33 feet). Above 11 meters (35 feet), extension ladders should have stay poles, called *tormentors,* that support the beams and give extra stability when raising or lowering the ladder. These ladders are often called Bangor ladders.

Roof or hook ladders come equipped with a folding hook at the upper end of each beam. When stowed, the spring-lock socket for the hook allows the hook to fold down out of the way; when released, the hook extends above the beam. These ladders are designed for working on peaked roofs. The ladder is laid flat on the roof and the hooks extend over the ridgepole, digging into the far side of the roof for stability and support. Fireboats use hook ladders, usually with fixed hooks, for boarding other vessels or to climb onto piers and wharves.

At times, ladders are needed inside a building. But because of the limits imposed by stairway and door clearances, ordinary straight ladders of the size needed cannot be used. Folding ladders that reduce to very short length were developed to meet this problem. Their main drawback is that they can safely support only one person at a time.

FIGURE 5-9. Thirty-five-foot trussed beam extension ladder in use. Courtesy of the Boulder, Colorado, Fire Department.

Rarely used, but still carried occasionally on ladder trucks, are scaling, or pompier, ladders. These are little more than a metal pole, with a long, slightly curved beak at the top and short projections for ladder rungs. The toothed beak will get a firm purchase on window sills, cornices, and other structural features, allowing the user to climb up or down, one story at a time, remove the hook and repeat the process as needed. Scaling ladders find their best use in rescue where a ladder already in place falls short of the endangered person.

NFPA Standard No. 1931* calls for testing at any time a ladder is suspected of being unsafe, and in any case at least once a year. Careful maintenance will add many years of useful life to fire ladders, particularly those of wood. Dirt, grit, oil, and grease can endanger life by speeding wear and creating dangerous conditions. Ladders should therefore be inspected after every use and carefully cleaned. Another important safety measure is checking the brackets and locks that hold the ladders flush with the sides of the apparatus. Failure of one of these while on the way to a fire could mean death or serious injury to any fire fighters on the side of the truck or to citizens on the street.

Aerial Equipment

The greatest advance in ladders was the development of the aerial ladder, mounted on a turntable, capable of extending up to 30.5 meters (100 feet). The San Francisco Fire Department acquired the first ladder of this kind in 1872, a pioneer model that used a worm gear, hand cranked, to raise and lower the ladder. Next came compressed air in pistons, followed by spring-assisted systems. The modern form, shown in Figure 5-10, uses a completely hydraulic system to raise and lower the ladder and move the turntable.

Latest models of the metal aerial ladder have three or four sections of ladder, depending on overall length and manufacturer, that are raised and lowered by hydraulically controlled cables. There is a stabilizing jack on each side to prevent overloading of the truck's rear springs; the jacks also prevent sag on the side toward which the aerial extends and hold the turntable platform level.

While the aerial ladder offers an almost miraculous assist to rescue work, it is of equal value in getting water on a fire in upper stories. Standard ladders come equipped with a ladder pipe that will allow the use of a heavy stream, if desired, on the upper floors of buildings; some models carry two ladder pipes. By connecting to the ladder pipe, those fighting fires on upper floors can easily charge their lines.

*National Fire Protection Association, *Standard on Fire Department Ground Ladders,* 1975 ed. (Boston).

FIGURE 5-10. Hydraulic aerial ladder. Courtesy of Pirsch, Inc.

Ladder trucks have another important function in fire fighting: carrying an assortment of tools and equipment every pumper does not need and could not carry, in addition to its hose load. The NFPA list of basic and required items includes that darling of motion pictures, the folding circular life net, and one or more of such things as rubber insulated gloves for electrical work, 18-tonne (20-ton) hydraulic jacks, blankets, inhalators or resuscitators, door opener, crowbar, and gas masks. Desirable items add another group of items to this list. (For details, see NFPA Fire Protection Handbook, Section 9, pp. 54–55.)*

When we add up these standards of equipment, we can easily see that a modern ladder truck is equipped to do many things. But to make full use of its capabilities, a ladder company needs more crew members than have been customarily assigned in the past. The ISO grading requires six people be assigned, with increasingly severe deficiencies assigned for each person fewer than the standard. Employee costs are seriously affecting crew levels in this category of apparatus, particularly in the smaller community because the number of fires there requiring the use of aerial trucks is low. Communities often assign one or two fire fighters to this six-person unit, which obviously reduces the overall fireground efficiency of the company.

*National Fire Protection Association, *Fire Protection Handbook*, 14th ed. (Boston).

Combinations

Because apparatus that meets the ISO Standard Schedule's requirements and NFPA Standard No. 1901 is so costly and trained people are in such short supply, many smaller fire departments resort to "combinations."

One widely used arrangement is to mount a 568 lpm (150 gpm) booster pump, a 1140- to 1890-liter (300- to 500-gallon) water tank, and a reel carrying from 61.0 to 122 meters (200 to 400 feet) of small hose on an aerial ladder truck. Another common form of combination, familiarly known as a *quad*, consists of a pumper-ladder combination that carries all the equipment of a pumper and a ladder truck, plus a water tank that may hold as much as 3785 liters (1000 gallons). Tanks over 2840 liter (750 gallon) capacity tend to cancel out cost savings since they need an extra-strength chassis to carry them. When a power-operated aerial ladder is added to a quad, it becomes a quintuple (fivefold) combination, known as a *quint*. To meet NFPA standards, as well as those in the ISO Schedule, combination trucks must carry the equipment listed for both classes of service. When an item appears on the equipment list for both types of apparatus, the largest number of any item specified is what must be carried.

Elevating Platform Apparatus

One of the few pieces of equipment that was not developed specifically for use by the fire service, the elevating platform in use today is a modified and improved version of the "cherry picker" used by utility companies. There is controversy as to who actually used the first elevating platform in the fire service. Today's elevating platform may use an articulating (jointed) boom or may be mounted on a telescoping boom. Figure 5-11 shows the articulated boom. (See *Fire Fighting Apparatus and Procedures*, by Lawrence W. Erven, for more detail on this type of aerial apparatus.*)

Operated by a hydraulic system, elevating platforms can reach from 20 to 27 meters (65 to 90 feet) above ground level. The platform or basket, with a load limit of about 363 kg (800 lb) can be controlled either from the ground, at the turntable, or by an operator in the basket. Safety devices make it possible to lower the platform in the event that the hydraulic system fails.

Since the introduction of the elevating platform, there has been much debate as to whether aerial ladders or snorkels are of more value in

*3rd Edition, 1979, Glencoe Publishing Company, Encino, California.

FIGURE 5-11. Elevating platform (articulated boom). Courtesy of the Boulder, Colorado, Fire Department.

fire fighting operations. Each has certain advantages and each has certain disadvantages.

Comparing the two types of apparatus, the snorkel has a definite advantage as a source of heavy streams. Its water supply, connected to a permanently installed turret nozzle on the platform itself, comes from three gated 2½-inch hose inlets (or other appropriate hose coupling). The turret has a minimum capacity of 1890 lpm (500 gpm); some can throw more than 3785 lpm (1000 gpm) in either a solid stream or a fog pattern. Also, adding two 2½-inch outlets at the platform gives a ready source of water for lead lines into the upper floor of a building. More important, the turret on a snorkel can be swept horizontally as well as vertically and can be controlled more closely than the stream from an aerial ladder pipe.

Aerial ladders have a three-way inlet connection for 3-inch or larger hose that feeds a heavy stream nozzle at the top of the bed section of the ladder or the tip of the top section, the fly, or both. The largest size standard nozzle mounted is 1¾ inches, giving a maximum of 3080 lpm (813 gpm) at 5.50 kPa (80 psi) and 3440 lpm (909 gpm) at 6.89 kPa (100 psi). To supply a nozzle of this size requires two hose lines instead of a single 3- or 3½-inch connection. Even then, maximum flow and pressure will be less than for the snorkel.

While icing in subfreezing weather is a problem for both types of apparatus, the snorkel platform and boom will accommodate built-in heaters, a modification still in the experimental stage but one that cannot be extended to aerial ladders. Notice, though, that an aerial ladder can be placed and put into operation faster than a snorkel because the stabilizing system for an elevating platform is much heavier, takes more room, must be operated by a hydraulic system, and the boom will not elevate (in the latest models) until the outriggers are fully extended and locked.

Rescue operations offer the best example of the difference between the two pieces of apparatus. The snorkel's best use is to lower injured people, particularly litter cases, or to sweep across a row of windows in a fire building, picking up one or two persons from each opening. Where a single opening offers an escape route to a number of people, an aerial ladder can evacuate them far more rapidly since it affords a continuous strong bridge to the ground. Raising and lowering an elevating platform is slow work. The snorkel also can be used as a mount for block and tackle or other rescue equipment to get people out of cave-ins, excavations, or the bottoms of tanks, while an aerial ladder is not suited for this purpose.

Ventilation operations give the snorkel a big edge. The platform can sweep across a building front using its turret's heavy stream or men with axes to break out windows as needed; each window ventilated requires separate raising, lowering, and shifting of an aerial ladder. Roof ventilation can often be handled from the platform of the snorkel and the stable platform makes handling power tools, such as saws or drills or jack hammers, easy. By comparison, using such tools from an aerial ladder is difficult.

Another advantage of the elevating platform is that it can be used as an elevator to lift hand lines, hand tools, and short ladders up to its load limits. This is helpful in overhaul and salvage work. And some chiefs have been known to use the platform as a reconnaissance platform or a place from which to direct operations.

To round out the picture, we must take into account the aerial's much longer horizontal reach at both upper and lower elevating angles, the fact that an aerial ladder can be raised much faster than a snorkel and that it can safely be placed 11 m (35 feet) or more away from a fire building. An elevating platform may have to go in as close as 7.3 m (24 feet) with consequently greater exposure to heat, flame, and smoke. In addition, the long booms of some snorkel models are hard to use in narrow streets or where there are overhead wires.

In spite of comparisons, the point is that both kinds of apparatus can be used together; each has its place in the fire service.

Heavy Stream Devices

When we face a big fire that calls for reach and penetration of great quantities of water, we need to use heavy streams, sometimes called *master streams*. As we have just seen, elevating platforms have fixed turrets that deliver a heavy stream and aerial ladder pipes accomplish the same thing by using large diameter hose nozzles. Fireboats and special land apparatus have nozzles up to 10 cm in diameter (4 inches), delivering 3785 lpm (1000 gpm) of water at pressures up to 620 kPa (90 psi).

People cannot handle such tremendous streams. The human limit is about a 2½-inch line, putting out 1510 lpm (400 gpm), and even this is risky enough to make a nozzle holder desirable. Above this limit, far below heavy stream volume and pressures, fire fighters must resort to portable turrets, turrets on apparatus, nozzle assemblies on a rigid mount, or water towers. Figure 5-12 shows a portable turret of the type commonly carried on pumpers. The features of this heavy stream appliance are common to all turrets, whether portable or mounted on apparatus. Three lines, at least 2½ inches in diameter, deliver water to the turret chamber; two curved pipes, with no sharp bends, carry the water to the single barrel of the turret for delivery on the fire. Tips of different sizes, from 1½ inches up to 2 inches or more, and large capacity fog nozzles, are interchangeable. Elevation and rotation can be controlled by hand, by

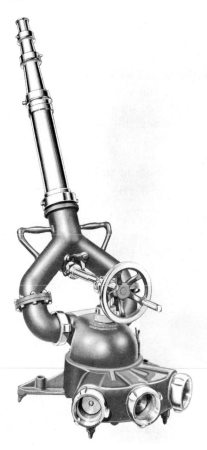

FIGURE 5-12. Example of a portable turret nozzle. Courtesy of the Akron Brass Company.

a worm and gear, or by both means. An important safety feature is an elevating lock that holds the nozzle at the minimum angle needed to keep the turret stationary. This must be used because if the nozzle is left at a lower angle, thrust under pressure may cause the entire turret to move backwards and whip out of control.

Mounted on the chassis of a vehicle, a fixed turret can handle heavier streams than a portable turret and takes less time to set up and put into operation. It also offers the advantage of having the stream start about 2.7 m (9.0 feet) above ground level, extending maximum vertical reach of an effective stream.

Another form of heavy stream appliance, permanently mounted, is the hydraulically operated giant deluge gun. Fed by multiple intakes that will take either 2½- or 3½-inch hose, this piece of equipment can deliver 11360 lpm (3000 gpm) at pressures from 210 to 2070 kPa (30 to 300 psi) and has an effective range of more than 97.5 m (320 feet). Tips vary from 1½ to 3 inches. Design differs from ordinary turrets by having a helical barrel and a 270-degree loop that balances all vector forces and thus allows a single pipe feed to the barrel instead of the turret system of splitting and then reuniting the stream.

We have already discussed the use of ladder pipes as heavy stream devices. One additional comment is in order: hydraulic and electrical control systems are now available that allow an operator at the turntable to raise or lower the nozzle and to change the nozzle pattern at will. On these models, the ladder pipe has a 3½-inch waterway with ball joint connections.

A few large cities still use water tower trucks for heavy stream operations, particularly on upper stories of buildings. (See Figure 5-13.) The apparatus consists of a portable standpipe, supported by a rigid tower mast, with one or more large, fixed nozzles at the end of the pipe. Maximum height varies from 17 to 23 m (55 to 75 feet).

FIGURE 5-13. Seventy-one-foot water tower. Courtesy of the San Francisco Fire Department. Photo by Chet Born

Although the American Insurance Association calls for either water towers or ladder pipes in high value districts that have more than ten buildings six stories high or taller, the water tower is gradually vanishing. Seventeen large cities have scrapped theirs. The Detroit Fire Department, in a report explaining why it had no water tower, had this to say:

> We junked our water tower in Detroit about 10 years ago and have since relied on ladder pipes for elevated streams. We have experienced no difficulty. The ladder pipes have a distinct advantage because of their versatility and the fact that they are on the job at the start. All Detroit ladders are of the aerial type and all are equipped with ladder pipes.

The elevating platform, as we have seen, also tends to cancel out the water tower, offering a much more flexible and maneuverable stream at the same or greater heights.

Other Vehicles

Salvage Trucks

One of the chief goals of a fire department is to protect property from destruction by fire. Back in the days before public fire protection became the rule, private fire companies sold protection to shopkeepers, business people, and the owners of buildings. Each company affixed its own distinctive "mark" to the buildings of its clientele signifying that these structures were under their protection. Unfortunately, competition was keen and when a fire broke out, rival companies would race to the scene and go into action, ignoring all marks. Frequently, some fire historians claim, during fist fights to see who would fight a given fire, the building burned to the ground and its contents were destroyed. As a way to prevent property loss, this left something to be desired.

During the same period, and even down to comparatively recent times, fire fighting techniques were crude. Windows were smashed, doors that were actually open were broken down, furniture or goods thrown out of windows onto sidewalks or wet ground, and holes chopped in walls and ceilings with little regard to actual need. These practices formed the foundation of the image of fire fighters as reckless vandals, waving axes and wrecking everything in sight. More important, excessive use of water often left a home or business with soaked goods, ruined plaster, pools on the floors, and dripping ceilings. The cure was more painful than the disease: very often, when a fire was out, the owner of the premises or the tenant was ruined.

The first group to take action to correct this sorry situation was the fire insurance companies. Faced with high damage claims, they formed their own salvage companies to cut their losses. These privately operated companies responded to all fire calls. While the fire fighters worked to put out the fire, the salvage company workers rescued property from damage by smoke, water, falling debris, and rough handling.

What happened was that fire departments, both large and small, came to realize that a high rate of loss due to fire had an adverse effect on the public, on insurance rates, and on the money that was available for organized fire protection. As a result, departments began to train people in salvage techniques and to provide them with the equipment needed to do this work well. As a public relations approach, the introduction of salvage considerations into fire fighting has had no equal. No longer does the public fear the ravages of careless fire fighters. Instead, homes and businesses are often left cleaner after a fire than before. People are grateful for the consideration shown them. Damage is cut by using the right amount of water to put out the fire, instead of flooding the place. Using salvage covers, opening doors and windows instead of smashing them, opening attic vents rather than cutting a hole in the roof, all help improve the fire service's image.

The need to consider property has had a beneficial side effect: it has forced fire fighters to make more efficient use of their equipment. Improved hose nozzles and other equipment, plus techniques that restrict damage, are the result of this concern; so, too, is increased research into better ways to fight fire, with less consequent damage.

Most fire departments charge all fire companies with preventing damage from heat, smoke, water, and other sources such as forceable entry and ventilation practices. This salvage role should enter into all operations at a fire, by everyone concerned. However, the role of salvage is often allocated to the truck company.

Special salvage squads are an advantage, however, and some cities still maintain individual salvage companies. In recent years, the Los Angeles Fire Department replaced many of its salvage trucks with a combined salvage-rescue unit. The Palo Alto, California, Fire Department developed a special salvage unit, but combined it with the responsibilities of a truck company, increasing both the number of people and the equipment available at the scene, and, importantly, the possibility of simultaneous salvage work being done during an emergency. (See Figure 5-14.)

The Texas Fire Insurance Department once set a standard of one salvage company in cities of 200,000 population, plus one additional company for every 200,000 more persons. This means that a city of 600,000 would have three salvage companies. The salvage company is not dead; a number of chiefs favor such companies but are unable to have one because of current personnel shortages. Since the chances of a revival of this specialized company seem reasonably good, let us consider what equipment the salvage company needs and uses.

FIGURE 5-14. Palo Alto Fire Department's special salvage truck.
Courtesy of the Palo Alto Fire Department.

The first and most important item of a salvage truck is a large number of salvage covers. These are generally made of heavy, waterproofed canvas, although several types of plastic covers have met with some success. There are also several kinds of treated paper covers, commonly used in building construction, available in large or small rolls. The standard canvas covers measure 3.7 × 5.5 m (12 × 18 feet), or 4.3 × 5.5 m (14 × 18 feet) and have an average weight of 13.6 kg (30 lb). Like most tarpaulins, they come with metal grommets spaced at intervals around their edges. Since the NFPA recommends between 40 and 100 such covers on a salvage truck, we can easily see why the average ladder truck cannot meet this requirement. The weight alone would be prohibitive.

The NFPA also lists approximately 50 essential items for salvage operations. Included in the list are two ladders 2.4 to 3.0 m (8.0 to 10 feet) long, bags of sawdust, a canvas water chute, wooden lath, hammers, nails, tarpaper, heavy duty stapler, floor runners to prevent tracking mud and dirt, wood and steel shovels, mops and wringers, pick-up bag (eight feet square with rope handles), S-hooks for securing coverings, sponges and chamois, water eductors for basement drainage, six 3.785-liter (four-quart) pails, sprinkler head stops, replacement units and wrenches to do the job, smoke ejectors, hand and power saws, pipe plugs and caps to keep down damage from ruptured water, steam, chemical or sewer pipes, portable electric generators, and hand lights.

Squad and Rescue Trucks

Squad trucks are vehicles that supplement the response of regular pumper and ladder companies. They carry breathing apparatus, protective clothing, resuscitators or inhalators, first-aid, and rescue equipment.

Many are also equipped as initial attack pumpers, apparatus that goes directly to the fire building while other pumpers lay large hose lines and connect to hydrants. The crews of squad trucks, men trained in emergency work as well as general fire fighting, help make up personnel deficiencies in regular companies and add to the force available in busy districts.

Rescue trucks are equipped with a large assortment of tools to carry out their special functions. Recommended minimum crew is six fire fighters and an officer. The work of such a company calls for trained specialists since any call may require them to perform such jobs as shoring, rigging, cutting with oxyacetylene torches, jacking up trucks and trains, and so forth.

Some rescue companies can handle such jobs as cutting heavy dock planking, timbers, and double flooring, breaching walls and concrete flooring, cutting bars and metal obstructions at fires, making openings in ships' sides, or untangling the results of vehicular or train accidents. The rescue company also carries devices for hoisting, bracing, wedging, and similar jobs. Refrigerator tools, rubberized protective suits, self-contained demand type masks, smoke ejectors, and special extinguishing agents extend its range of operations. Rescue companies often carry medical supplies, stretchers (Stokes basket stretchers), blankets, splints, resuscitators, and so forth. Many of these units are now being designed as separate ambulance or paramedic units that handle only emergency medical calls.

Fireboats

For fire fighters of nautical bent, coastal cities and ports offer a special attraction: service aboard a fireboat. Waterfront areas usually represent a high fire hazard because of substandard construction (a carryover from the days when they were built), large amounts of combustible materials on hand, and limited access from the land side. In addition, a wharf or pier fire frequently gets into the space between the surface of the wharf and the water where it finds ample fuel in the form of pilings, timbers, cross bracing, and floating oil and grease. Access, even from the water side, is extremely difficult.

Experience shows that fireboats are not too well adapted to fighting pier warehouse or storage fires, unless the goods are piled in the open air. Even the heavy streams that a fireboat's turrets can throw are not enough to contain a fire in long pier structures that lack partition fire walls, or draft curtains and ventilation skylights. Their best use is in fighting fires aboard other vessels. Here the powerful streams can extinguish cargo fires or provide a protective curtain for the fireboat while it tows a burning ship away from wharves and exposed structures. Fireboats are also able to intercept a burning vessel while it is still at sea and prevent its total loss.

In addition, a fireboat can be very useful in pumping out vessels that are in danger of sinking and as a pumper supplying water to land-based engine companies. Many fireboats have rated pumping capacities in excess of 37 850 lpm (10 000 gpm) at 1030 kPa (150 psi) pressure.

Fireboats come in many sizes, from the big seagoing tug that may cost one million dollars, down to small, high-speed boats carrying a portable pump and small lines for work under the wharf. The size found in any given port depends on the conditions in the harbor that fix length, width, and draft of the boat, as well as the kind of service the fireboat is expected to perform.

A fireboat of modern design (see Figure 5-15) is 20 m (65 feet) long, has a beam (width) of 5.6 m (18½ feet), draws 1.8 m (6.0 feet) of water and can travel at about 24 km (15 miles) per hour. Two 600-hp (447 kW) diesels drive its two pumps, delivering 13 250 lpm (3500 gpm) each and two smaller diesels, 115 hp (85.8 kW) each, provide the power needed to hold position when not at anchor or docked. Four turret nozzles on the deck and a turret nozzle on a tower make up the heavy stream equipment; ten outlets on the deck can supply 2½-inch hose. The boat is equipped for towing and carries both radar and radio communication equipment.*

FIGURE 5-15. Fire Boat—City of Oakland. Courtesy of the Oakland Fire Department.

*Refer to *Fire Company Apparatus and Procedures*, by Lawrence W. Erven, Figure 2-11, Jet-Powered Fireboat (p. 26), for additional information on fireboats. 2nd ed., 1974; Glencoe Press, Beverly Hills, California.

NFPA-recommended equipment includes at least 304.8 m (1000 feet) of 2½-inch hose, plus 91.4 m (300 feet) of 1½-inch hose, hand tools such as axes and crowbars, short ladders, self-contained breathing apparatus, cellar, distributing, and large capacity fog nozzles, siphoning equipment with strainer, and a water eductor. A fireboat should also have a fixed foam system and systems for delivering carbon dioxide or dry chemicals, when needed.

Some cities, such as Los Angeles, have trained underwater teams that are called into service when hose streams have to be directed under birdges, piers, or wharfs. In certain other cities, such as Denver, these crews also recover the bodies of people who have drowned.

The ISO grading schedule requires a fireboat where occupied wharf frontage exists totaling 1.6 km (1.0 mile) that has buildings, storage, and so forth that would require fire fighting operations from the water side. Such wharf frontage should not be further than 2.4 km (1½ miles) from a fireboat. Each such fireboat should be able to deliver at least 18 920 lpm (5000 gpm) or one-half the required fire flow for the area protected.

Staffing fireboats can be a very expensive proposition. Large vessels are subject to federal marine regulations and must carry qualified pilots and crews as well as the fire fighting company. Fire fighters needed include an officer, an engineer, one member for each turret, and several members for the hand lines on the deck. There should, of course, be a reserve of trained personnel as well as a complete second or third shift.

Smaller fireboats do not need such large crews, are far less costly, and, if able to deliver from 1890.5 to 3785 lpm (500 to 1000 gpm), may well serve to protect small harbors and marinas.

Power and Light Trucks

Power and light trucks are specialized equipment (Figure 5-16) consisting of a vehicle with a large capacity electric generator, floodlights and spotlights (both portable and fixed), heavy electrical cable, and an assortment of electrical power tools. Fire departments find it safer and more practical to light up both inside and outside areas of fire sites at night. Fires may require all four sides and the roof to be illuminated, plus lights on the interior. One expensive large light unit is usually insufficient, and only the larger departments might be able to afford more and staff them. Simultaneous fires also require additional light units and the problem grows. Modern practice calls for generators, transformers and service outlets on all fire apparatus, supplying current to all floodlights and power equipment carried by the particular unit. Lightweight floodlights, using high intensity bulbs, are quite effective, yet

FIGURE 5-16. Lighting unit. Courtesy of the San Francisco Fire Department. Photo by Chet Born

less costly overall. See Palo Alto Fire Department's Salvage Unit (Figure 5-14) for one example of lightweight, portable lights that are often used on other fire department vehicles in smaller departments to accomplish lighting at fires.

Fire Extinguishers

NFPA standards require at least two approved portable extinguishers on each piece of fire apparatus. The usual complement consists of a

9.5 liter (2½ gallon) foam extinguisher and a 9 km (20 lb) dry chemical extinguisher for use on Class B fires, which are those occurring in flammable petroleum products or other flammable liquids or greases where ordinary portable extinguishers and water are of no use. Their use may allow a quick knockdown of such fires or containment until heavy equipment can be brought into play.

Protective Equipment

A serious hazard present in many fires is the toxic potential of the materials involved as well as normal combustion products. Earlier chapters only touched on the dangers of poisonous gases, corrosives, and substances that poison on contact with the human body. Yet, in our highly industrialized society, these dangers often exist. The following list of industrial materials contains only six examples of the many toxic substances that are manufactured, shipped, stored, and used by the ton, every year:

CHEMICAL	ANNUAL U.S. PRODUCTION IN TONS (APPROXIMATE)
Sulfuric Acid	28 million
Ammonia (anhydrous)	12 million
Chlorine	8 million
Sodium Hydroxide (lye)	8 million
Nitric Acid	7 million
Phenol	300,000

While only two of these industrial chemicals, ammonia and phenol, are flammable, all are dangerous to health and life. To deal with the dangers posed by any industrial material involved in a fire or leaking from a container, fire fighters must be able to identify the substance quickly and accurately. Some dangerous chemicals, such as cyanide and arsenic compounds, carry a built-in danger signal, but there are tens of thousands of other chemical compounds used in modern industry whose names mean nothing to the average person. When confronted with any unknown chemical, the best practice is to assume it is toxic and proceed accordingly. We will discuss the proper measures later in this chapter.

Fire fighters can get some help in identifying what kind of substance is present from labels, placards, and color coding systems on containers. Warning labels of the Manufacturing Chemists Association (MCA) carry a brief description of any poison, recommended procedures to follow if it is spilled or involved in a fire, and suggested antidotes. Some departments and companies keep a file of MCA "Chem Cards" (see Figure 5-17) that give information on a limited number of chemcials.

MCA CHEM-CARD — Transportation Emergency Guide

> **CC-48**
> February 1978

CARBOLIC ACID
(Phenol)

Clear or pink liquid which darkens on exposure to light; sweet odor

 IMMEDIATE HAZARDS

FIRE *Can catch fire.*

EXPOSURE *SPEED IN REMOVAL FROM SKIN IS OF UTMOST IMPORTANCE. Poisonous by skin absorption; may be fatal. Causes severe burns.*

IN CASE OF ACCIDENT

IF THIS HAPPENS

For assistance, phone
CHEMTREC
toll free, day or night
800-424-9300

DO THIS

SPILL or LEAK

Keep people away. Shut off leak if without risk. If necessary to enter spill area, wear self-contained breathing apparatus and full protective clothing. Dike large spill, allow to solidify or, if possible, pump into salvage tanks. Avoid flushing area with water, spill area should be neutralized. Run-off to sewer/waterway may create toxic hazard. Notify authorities of possible pollution & fish kill.

FIRE

On small fire use dry chemical or carbon dioxide. On large fire use water spray or foam. Wear self-contained breathing apparatus and full protective clothing. Cool tank with water if exposed to fire.

EXPOSURE

Speed in removal from skin is of utmost importance. Scrub skin with water and cloth pad. Remove contaminated clothing and shoes at once. In case of eye contact, immediately flush with plenty of water for at least 15 minutes. Use artificial respiration if not breathing; use external cardiac massage if definite cardiac arrest. Call a physician. Keep patient at rest.

FIGURE 5-17. Typical Chem-Card. Copyright 1971 by Manufacturing Chemists' Association, Inc. Reprinted by permission.

NFPA Standard No. 704, *Fire Hazards of Materials,* visually illustrates the hazards to fire fighters fighting fires in fixed installations such as chemical processing areas, storage and warehouse facilities, and laboratory entrances. The "704 diamond" uses colored numbers to illustrate the degree of hazard in health and the flammability and reactivity (Figure 5-18).

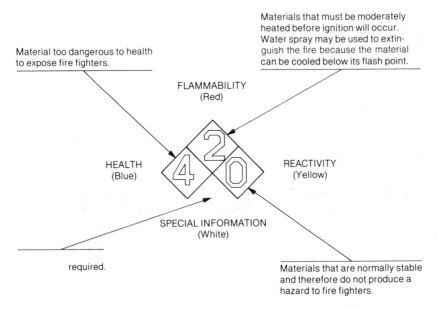

Material too dangerous to health to expose fire fighters.

Materials that must be moderately heated before ignition will occur. Water spray may be used to extinguish the fire because the material can be cooled below its flash point.

FLAMMABILITY (Red)

HEALTH (Blue)

REACTIVITY (Yellow)

SPECIAL INFORMATION (White)

required.

Materials that are normally stable and therefore do not produce a hazard to fire fighters.

FIGURE 5-18. 704 Diamond.

The higher the number (0–4), the greater the degree of hazard. These numbers and symbols are easily recognized by fire fighters and they can take immediate steps to protect themselves. Unfortunately, this method has not been widely adopted by the general public or governmental units other than fire departments.

The Department of Transportation (DOT) labeling system is perhaps the most widely used. New, more descriptive placards and labels have been selected that permit fire fighters and emergency service people all over the world to visually understand the hazards they face at the scene of an emergency. Some of the more common labels and placards are described here (see Figures 5-19a and 5-19b), and should be explored in depth.

COLOR SCHEME FOR NEW PLACARDS

Placard	Colors
DANGEROUS	Upper and lower triangles in red; inscription in black on white
EXPLOSIVES A and EXPLOSIVES B	Orange background; symbol and inscription in black
NON-FLAMMABLE GAS	Green background; symbol and inscription in white
OXYGEN, OXIDIZER and ORGANIC PEROXIDE	Yellow background; symbol and inscription in black
POISON GAS, POISON and CHLORINE	White background; symbol, borderline, and inscription in black
FLAMMABLE and COMBUSTIBLE	Red background; symbol and inscription in white
FLAMMABLE SOLID	White background with seven vertical red stripes; symbol and inscription in black
FLAMMABLE SOLID W	Triangle at top blue with white symbol; rest of placard white with seven vertical red stripes and inscription in black
RADIOACTIVE	Top portion yellow with black symbol; lower portion white with black inscription
CORROSIVE	Center and lower areas black; inscription in white; symbol in black and white

NOTE: The word *"Gasoline"* may be used in place of *"Flammable"* for highway transportation of gasoline. The words *"Fuel Oil"* may be used in place of *"Combustible"* for highway transportation of fuel oil that is not classed as a *"flammable liquid."*

IMPORTANT

1. All four sides of the vehicle must be placarded. Placement of the front placard may occur on either the cab or the cargo body.
2. Placard must be placed at least 3 inches away from any other marking or sign. Double placarding should be adjacent to each other.
3. Combinations of vehicles, each of which contains hazardous materials, shall each be placarded in accordance with the above chart.

FIGURE 5-19a. Hazardous material identification placards.

HAZARDOUS MATERIALS PLACARDING REQUIREMENTS
(ANY QUANTITY MUST BE PLACARDED)

HAZARD	NEW PLACARD
Class A Explosives	EXPLOSIVES A
Class B Explosives	EXPLOSIVES B
Poisons—Class A	POISON GAS
Flammable Solids (With Dangerous When Wet Label on Container only)	FLAMMABLE SOLID W
Radioactive Materials (Radioactive III yellow container label)	RADIOACTIVE
Radioactive Materials (Uranium hexa-fluoride, fissile)	RADIOACTIVE & CORROSIVE

PLACARDING MIXED LOADS

The **DANGEROUS** placard must be used for mixed loads containing more than one kind of hazardous material requiring placards when the aggregate gross weight totals 1000 pounds or more, except the following which require their own specific placard **FOR ANY QUANTITY.**

EXCEPTION A

Explosive A
Explosive B
Radioactive (Yellow III)

Poison Gas (Class A)
Flammable Solid (Dangerous When Wet)

EXCEPTION B

Specific placards are also required for each of the other classifications **if 5000 pounds or more** (gross aggregate weight) of one class of such material is loaded at one loading facility.

DOUBLE PLACARDING should be used when loads requiring **DANGEROUS** placards are mixed with any quantity of the classes shown in Exception A, and with 5000 pounds or more of all other classes as described in Exception B.

HAZARDOUS MATERIALS PLACARDING REQUIREMENTS
HIGHWAY—PLACARDS FOR MINIMUM REQUIRED QUANTITY ONLY
(NORMALLY QUANTITIES OVER 1000 POUNDS)
RAIL—PLACARD FOR ANY QUANTITY

HAZARD	NEW PLACARD
Class C Explosives	FLAMMABLE
Non-flammable Gas	NON-FLAMMABLE GAS
Non-Flammable Gas —Chlorine	CHLORINE
Non-Flammable Gas—Oxygen	OXYGEN
Combustible Liquids—Packages with rated capacity of 110 gallons or more, cargo tanks or tank car	COMBUSTIBLE
Flammable Liquid	FLAMMABLE
Flammable Solid	FLAMMABLE SOLID
Oxidizer	OXIDIZER
Organic Peroxide	ORGANIC PEROXIDE
Class B Poison	POISON
Corrosive Material	CORROSIVES
Irritating Material	DANGEROUS

FIGURE 5-19b. Hazardous material identification placards.

The U.S. Department of Transportation has also published an *Emergency Action Guide for Hazardous Materials.** This guide is intended to assist the emergency services personnel during the first 30 minutes of an incident involving known volatile, toxic, gaseous or flammable material shipped in bulk. See Figures 5-20a and 5-20b for examples of the various kinds of emergency information available. The placarding required for the transportation of chlorine is shown at the top of Figure 5-20a.

All of these methods of identification suffer from an unavoidable shortcoming: they are likely to be invisible in the gloom and smoke of a fire or burned beyond recognition. As a result, positive identification of an industrial material may be extremely difficult in some emergency situations. This is why prefire inspection of occupancies that store or use these substances is so valuable. Preplanning permits a complete and accurate survey of the hazards and provides sufficient time for a thorough discussion of alternative tactics. On the highway, preplanning is less practicable, except in a generalized manner. Quick identification may then hinge on gaining prompt possession of bills of lading or other shipping papers.

Assuming a substance can be identified as toxic or poisonous, what does this mean to a fire fighter? By definition, a poison is a substance that is injurious to health even when it is absorbed in minute amounts. Safely stored in a proper container, a poison is only potentially hazardous. But, when a poison escapes from its container, or is produced in some manner during an emergency, the routes of possible absorption into a person's body become significant. Wearing protective equipment is essential in order to prevent this absorption.

Respiratory Protective Equipment (Gas Masks)

By far the most common way for a fire fighter to become poisoned is by inhaling toxic gases or vapors. And, contrary to common belief, one's sense of smell gives little warning of danger. True, no one can miss the warning given by the sharp bite of ammonia or the choking stench of sulfur dioxide, but there are many far more dangerous gases that have a pleasing smell or none at all. For example, phosgene, a deadly gas, has an odor much like freshly-mown hay; carbon monoxide has only a very faint metallic odor; both can be breathed without any discomfort.

Poisoning by inhalation interferes with an essential process of the human body. When we inhale, we take in oxygen as part of the air

*Hazardous Materials—Emergency Action Guide, U.S. Department of Transportation, National Highway Traffic Safety Administration and Materials Transportation Bureau, Washington, D.C. Published May 1976; revised January 1977.

Chlorine
(Nonflammable Gas, Poisonous)

Potential Hazards

Fire: — Cannot catch fire.
— May ignite combustibles.

Explosion: — Container may explode due to heat of fire.

Health: — Contact may cause burns to skin or eyes.
— *Vapors may be fatal if inhaled*.
— Runoff may pollute water supply.

Immediate Action

— Get helper and notify local authorities.
— If possible, wear self-contained breathing apparatus and full protective clothing.
— Keep upwind and estimate *Immediate Danger Area*.
— Evacuate according to *Evacuation Table*.

Immediate Follow-up Action

Fire: — Move containers from fire area if without risk.
— Cool containers with water from *maximum distance* until well after fire is out.
— Do not get water inside containers.
— Do not use water on leaking container.
— Stay away from ends of tanks.

Spill or Leak: — Do not touch spilled liquid.
— Stop leak if without risk.
— Use water spray to reduce vapors.
— Isolate area until gas has dispersed.
— Do not get water inside containers.

First Aid: — Remove victim to fresh air. Call for emergency medical care. *Effects of contact or inhalation may be delayed*.
— If victim is not breathing, give artificial respiration. If breathing is difficult, give oxygen.
— If victim contacted material, immediately flush skin or eyes with running water *for at least 15 minutes*.
— Remove contaminated clothes.
— Keep victim warm and quiet.

Assistance Call Chemtrec toll free (800) 424-9300

In the District of Columbia, the Virgin Islands, Guam, Samoa, Puerto Rico and Alaska, call (202) 483-7616.

FIGURE 5-20a. Example of emergency action guide for hazardous materials.

Additional Follow-up Action

—For more detailed assistance in controlling the hazard, call Chemtrec (Chemical Transportation Emergency Center) toll free (808) 424-9300. You will be asked for the following information:

- Your location and phone number.
- Location of the accident.
- Name of product and shipper, if known.
- The color and number on any labels on the carrier or cargo.
- Weather conditions.
- Type of environment (populated, rural, business, etc.)
- Availability of water supply.

—Adjust evacuation area according to wind changes and observed effect on population.

Water Pollution Control

—Prevent runoff from fire control or dilution water from entering streams or drinking water supply. Dike for later disposal. Notify Coast Guard or Environmental Protection Agency of the situation through Chemtrec or your local authorities.

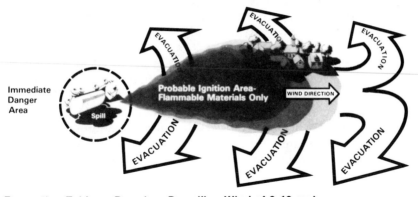

Evacuation Table — Based on Prevailing Wind of 6-12 mph.

Approximate Size of Spill	Distance to Evacuate From Immediate Danger Area	For Maximum Safety, Downwind Evacuation Area Should Be
200 square feet	160 yards (192 paces)	1 mile long, 1/2 mile wide
400 square feet	240 yards (288 paces)	1 1/2 miles long, 1 mile wide
600 square feet	300 yards (360 paces)	1 1/2 miles long, 1 mile wide
800 square feet	340 yards (408 paces)	2 miles long, 1 1/2 miles wide
In the event of an explosion, the minimum safe distance from flying fragments is 2,000 feet in all directions.		

FIGURE 5-20b. Example of emergency action guide for hazardous materials.

we breathe. Air passes through respiratory passages that lead to the lungs where pure oxygen is picked up by the red cells of the blood (hemoglobin) and transported by the blood stream to the cells of the body and brain. If this transfer is interrupted in any way for more than a very short time, we die. To prevent interruption, human beings exposed to poison gases must wear respiratory protective equipment, commonly called gas masks.

Canister Masks   One type of gas mask is the canister mask. Essentially filters, these masks protect the wearer against concentrations of smoke and low percentages of certain gases. They are useless where the oxygen content of the atmosphere drops below 16 percent. While this oxygen level will sustain life, any kind of physical effort is difficult and there is often a temporary but severe failure in judgment. Special canisters extend the usefulness of this type of mask. The red "universal" or "all service" canister, still used by many fire departments, gives only limited protection in spite of its name. The back of the canister warns:

"This canister is approved for respiratory protection in atmospheres containing 16% or more oxygen; and not more than 2% acid gases, organic vapors, or carbon monoxide; 3% ammonia; and 2% of poisonous gases if more than one class is present. Also approved for respiratory protection against dusts, fumes, mists, fogs, and smokes that are not significantly more toxic than lead."

Although a mask of this type is lightweight, easy to put on and use, easy to maintain, simple, and relatively low in cost, it affords only limited protection to its wearer. This is particularly true in the case of carbon monoxide, the toxic gas most commonly present at a fire.

Importance of Gas Masks

Any ordinary fire involving such common materials as wood, paper, grass, certain plastics, flammable liquids or gases, coal, charcoal, and other organic substances, produces both carbon dioxide and carbon monoxide. When a fire does not have enough oxygen supply, a situation quite common in enclosed spaces such as cellars or in piled or baled materials, incomplete combustion results in more carbon monoxide than carbon dioxide. Sometimes heavy black smoke, composed partly of unburned carbon particles, indicates incomplete combustion and the presence of carbon monoxide but this is not always the case. Burning natural gas and charcoal, for example, are materials that can create quantities of carbon monoxide with little smoke production.

Carbon monoxide poisons by "stealing" the oxygen in the blood. Blood has approximately 300 times the affinity for carbon monoxide that it has for oxygen, and when inhaled, carbon monoxide instantly combines with the hemoglobin of the blood. Loaded with CO, the blood cells can no longer transport the oxygen the body needs. Unconsciousness, caused by lack of oxygen in the brain cells, quickly follows and if the victim is not removed to fresh air and given oxygen soon after unconsciousness, death occurs. It also takes some time for the body to recover from carbon monoxide poisoning. Fire fighters (or others) overcome by CO should not be allowed to reenter a burning building or one charged with smoke because a small additional amount of the gas may cause a rapid return of the symptoms of poisoning.

Canister masks offer relatively little protection against carbon monoxide poisoning because if the concentration of CO found in the atmosphere of fire buildings exceeds the 2 percent protection limit of the mask, the surplus will simply be breathed in by the wearer. In fact, wearing a canister mask may be more dangerous than not wearing one at all. The mask will filter out the irritants commonly present in smoke or other fumes and thus allow the wearer to penetrate a dangerous atmosphere where he may be overcome.

Canister masks have other serious drawbacks. The Hopcalite catalyst, the most important part of the mask, cannot function when it is wet. Although there is an external valve to prevent moisture (including exhaled air) from entering the top of the canister and anhydrous calcium chloride protects the bottom, an atmosphere full of water spray may penetrate the drying agents. The timer, or indicator, that shows when the canister must be replaced or its two-hour-rated life is up may not be easy to see. Fire fighting experience shows that useful function often ends far sooner than two hours. Because the wearer must draw air through several layers of activating chemicals, breathing becomes harder. If toxic gases are present in percentages higher than the mask can handle, faster breathing merely speeds up the rate of intake. Similarly, when there are high concentrations of carbon dioxide (CO_2), breathing rate may increase to the point at which resulting oxygen demand cannot be met. Dizziness, irrational behavior, and collapse can follow.

While carbon dioxide and carbon monoxide are the most common gaseous products of combustion, there are several other gases and vapors commonly released during fires. Some of them are far more dangerous than carbon monoxide because inhaling only a small amount will prove fatal. Some give warning of their presence, some do not. Acrolein, liberated by burning fats or oils, is flammable, heavier than air, and has a disagreeable choking acrid odor. It is highly irritating to the nose, throat, eyes, and lungs.

Ammonia, usually found as an industrial gas or refrigerant, can be produced by burning wool or silk or when certain chemicals are heated. The gas is flammable in concentrations above 16 percent in air, highly irritating to the tender parts of the body and can cause swelling that blocks the throat and lung passages. Acid fumes, from heated or leaking containers, can have the same effect on the respiratory system. These fumes, as well as ammonia, warn of their presence by their sharp smell and other more serious effects on the body.

Hydrogen cyanide is sometimes released by the combustion of wool, silk, and certain plastics, liberated by the heating of cyanide and nitrite compounds and by the action of acids on cyanides. In a fire, the so-called bitter almond odor may be masked by irritants. Hydogen cyanide is flammable, water soluble, and extremely poisonous. Cyanide prevents the transfer of oxygen from the blood to the cells.

Hydrogen sulfide can be created when hair, wool, meat, hides, or chemicals that contain sulfur are burning. While very small amounts have a distinctive rotten egg smell, larger quantities quickly paralyze the sense of smell. It irritates the eyes, wet skin and breathing passages, causes swelling, and may quickly paralyze the respiratory centers. Sulfur dioxide, another combustion product of sulfur and sulfur compounds, is a nonflammable gas with an extremely irritating odor. It is largely used as a fumigant and in some older refrigeration systems. Inhalation results in swelling that blocks the throat and lung passages. Carbon disulfide can also paralyze the nerve centers that control breathing.

Still other poison gases or vapors attack essential organs or functions of the body. Vapors from Parathion and certain other insecticides short-circuit the transmission of nerve impulses. Benzene vapors, toxic at levels below those detectable by smell, can injure the bone marrow that produces red blood cells. Carbon tetrachloride and some other chlorinated hydrocarbons cause damage to the liver and kidneys. These vapors or gases are produced in various ways, including:

- Escape of a gas or vapor from a damaged container
- Reaction of a chemical with water or air
- Action of nitric acid or other chemicals on ordinary substances
- Interaction of different chemicals
- Heating of toxic liquids or solids
- Combustion of ordinary flammable materials

The nitrogen oxides, although sometimes visible as brown-to-orange gases, may be undetectable by either sight or smell in dangerous concentrations. These gases attack the lung tissue directly by decomposing in the moist passageways of the respiratory system to form nitric acid. The nitrogen oxides can be created by the heating or reactions of nitric acid, by heating of nitrates, such as certain commercial fertilizers, or when materials containing nitrogen, such as celluloid, other forms of cellulose nitrate, dynamite and other explosives, decompose or burn.

Similarly, with little immediate warning from pain, phosgene slowly decomposes in the lungs to form hydrochloric acid and carbon monoxide. Inhalation of even a very low concentration can be fatal. Phosgene can be formed accidentally when carbon tetrachloride or other chlorinated hydrocarbons are heated. It has been used as a chemical warfare agent.

Considering this range of toxic gases and the defects of the canister mask, it is small wonder that the fire service is rapidly changing over to self-contained breathing apparatus. Additionally, there are court decisions that tend to indicate that employers that require fire fighters to wear canister-type gas masks will be liable for injury or death caused by their use. These decisions have accelerated the rapid changeover to self-contained breathing equipment.*

Self-contained Breathing Apparatus

The lightest type of self-contained breathing apparatus is a canister or small metal container that generates oxygen once its seal is broken and the wearer has exhaled into the mask to start the reaction.

Another type is the oxygen rebreathing mask, good for one hour. This apparatus is complicated, having an oxygen cylinder, breathing bag, chemical cleaning elements, reducing valve, separate tubes for breathing and exhaling, and release valves. It must also be purged of nitrogen every 15 minutes by exhaling through the relief valve. In the hands of experienced rescue people, however, it serves very well.

Demand masks, now in wide use, resemble those masks commonly used for skin diving. A facepiece connects directly to a pressure-reducing valve that lets compressed air or oxygen out of a backpack tank or cylinder when the wearer inhales through the mouthpiece. Recharging the cylinders or substituting filled ones is quick and easy, but average useful time is no more than 30 minutes. In addition, this type of mask is rather heavy and bulky, making it hard to use while handling equipment or hose.

Demand masks are the most commonly used masks in most fire departments. Today, however, because of the increased toxicity of the atmosphere fire fighters may be exposed to, the positive pressure mask is also becoming popular. This unit has higher pressure than that of the atmosphere inside the facepiece, so that all gases and vapors are kept out of the mask should openings develop in the mask's seal.

Masks, like any other piece of equipment, require careful maintenance. They should be inspected after each use, or at least once a week. Any repairs needed should be made at once and defective masks replaced. Some states require every piece of apparatus to carry masks of one type or another.

*See *MacClave* v. *City of New York,* 24 AD2d 320, 265 NYS2d 222 (1965).

Various studies are being conducted on the design and safety aspects of self-contained breathing apparatus in use today. NASA has conducted a preliminary study that tends to indicate that "off-gases" given by rubber portions of some masks in hot atmospheres are often as toxic as those gases we are trying to be protected from. Extreme temperatures have shattered the clear plastic facepieces and have caused cracks in the rubber portion of the mask facepiece. Additional design studies are currently underway that should bring the older, heavy and awkward self-contained masks up to date with use of space age design and materials.

Skin and Body Protective Equipment

So far, we have discussed toxic substances that attack through the respiratory system. Unfortunately, there are other ways in which a poison can attack. Ingesting, or swallowing, a poison is a major cause of accidental death in the United States but a minor one in the fire service. This is because smoking or eating at the scene of an emergency where poisons or radioactive materials are present is strictly forbidden and the rules are enforced. In addition, wearing a proper mask prevents accidentally ingesting poisons through the automatic swallowing process that goes on all the time and is accelerated when facing danger or working hard.

Impermeable Clothing   To close off another route poisons take in attacking human beings, the fire service is increasing its use of impermeable clothing. Ordinary turnout clothing, even when a mask is worn, leaves the forehead, ears, and back of the neck and head exposed. Many fire fighters also prefer working without gloves, leaving their hands exposed. These are all avenues for attack by poisons that can be absorbed through the skin. As we saw earlier in this chapter, hydrogen cyanide is such a poison; so too are even tiny amounts of insecticides such as Parathion, TEPP, phosdrin, and other organic phosphate compounds. DDT and other chlorinated hydrocarbons are less toxic, requiring larger amounts to cause injury, let alone death. Phenol (carbolic acid) has caused industrial deaths when an external skin area not much larger than a person's forearm was soaked by the solution. Unlike corrosive substances, these and other skin penetrants give little warning of their presence. Knowing what is involved in a fire becomes doubly important if the fire fighter is to protect his own life and health.

Impermeable clothing is a special suit that covers the body completely and is made of a material that cannot be penetrated by gases, vapors, or liquids. When worn with a self-contained mask and impermeable gloves, it offers complete protection against poisoning by absorption through the skin. A field expedient that is sometimes used is to secure the legs and sleeves of turnout clothing with heavy rubber bands. This will prevent some types of gas and vapors from traveling up the legs and arms but it is far from safe.

Acid Suits   Acid suits serve to protect fire fighters against more than just acid and acid fumes. Made of such materials as Neoprene, they also serve to prevent the serious burns and tissue damage that corrosive substances can cause. While some solids and gases are similarly danger-ous to living tissue, the corrosive liquids are of greater concern to the fire service because they are among the most widely used of all industrial chemicals. Death can follow very rapidly, accompanied by great pain, if one of these substances is swallowed; breathing their fumes can cause severe internal damage to the respiratory system, and blindness is likely if they come in contact with the eyes. The value of inspections that discover their presence and preplanning on how to handle them cannot be overrated.

Corrosive poisons can also act directly on the external layers of the skin and literally eat their way down into the deeper tissues. Any contact with strong mineral acids, such as sulfuric, nitric, hydrochloric; the caus-tic alkalis, such as sodium or potassium hydroxide; or the other white label liquids, will generally cause severe pain and, if this warning is ignored, still more serious damage. A few, hydrofluoric acid among them, destroy nerves along with tissue, delaying the beginning of pain.

If an unprotected fire fighter is splashed by one of the major mineral acids, the damage can be greatly reduced by immediate washing by a hose stream or strong shower for at least 15 minutes. Contaminated clothing should be removed before starting to wash. Eyes exposed to corrosives should be irrigated with running water for the same length of time, and medical attention should be obtained while this first-aid treatment is taking place.

Even when fully protected by a self-contained mask and impermeable clothing, a fire fighter must not enter alone into an area where toxic materials are present. Personnel should work in teams and each team should be kept under close supervision by an officer.

Personal Equipment

Basic personal equipment issued to a fire fighter is usually a helmet, a waterproof coat, rubber boots (long or short), waterproof trousers, several sets of fatigues, and a uniform for street wear. Some departments provide many other smaller items connected with fire department work such as gloves, goggles, face shields, and special shoes. A very few pro-vide all necessary personal clothing except underwear and socks.

REVIEW QUESTIONS

1. What is a pumper and what is its purpose?

2. How does a centrifugal pump work and what are its advantages and disadvantages?

3. Describe the operation and use of:

   (a) auxiliary pumps

   (b) high pressure pumps

   (c) booster pumps

   (d) skid-mounted fire pumps

4. What is a triple combination pumper?

5. Describe a grass and brush fire truck and explain its use.

6. Describe an airport crash truck and explain its use.

7. What kind of hose is used in fire fighting and what are some precautions that should be taken in its care?

8. What are ground ladders? Describe their construction; what are their limitations?

9. Describe:

   (a) extension ladder

   (b) roof ladder

   (c) scaling or pompier ladder

10. What is an aerial ladder, how does it work, and what are its advantages and limitations?

11. Describe an elevating platform truck. Name its advantages and limitations.

12. Name two kinds of turret nozzles and explain how each is used.

13. Name other powerful stream appliances.

14. What is the value of a salvage truck in the fire department?

15. Name some of the equipment carried on a salvage truck and explain how it is used.

16. Describe and explain the use of squad and rescue trucks.

17. What are fireboats and when are they required?

18. What is light truck and what is it used for?

19. Name six hazardous chemicals from which fire fighters must be protected.

20. How can these chemical hazards be recognized?

21. Name the kinds of respiratory equipment and list advantages and limitations of each.

22. Name two types of clothing used to protect the skin from hazardous chemicals.

6 Extinguishing Agents

Water

While there are many specialized extinguishing agents, such as the spectacular airport crash foam blanket, *water* remains our first line of defense against fire and our primary extinguishing agent. This is because water is generally available in large quantities, at relatively low cost, and because of its remarkable properties of specific and latent heat. Water's extinguishing properties of cooling, smothering, emulsifying, and diluting are another reason that this particular agent is so useful in the variety of fire fighting situations. Furthermore, modern fire fighting techniques and equipment let us make better and far more effective use of water than ever before.

Heavy Streams

As we saw in Chapter 5, heavy streams, also referred to as master streams, are large volume, high pressure streams of water used to fight fires that are too large to combat with hand lines. Fires of this magnitude represent only $\frac{1}{3}$ of 1 percent of the annual fire total in the United States but they are the cause of more than 60 percent of our annual national property loss by fire.

Heavy stream appliances, such as we discussed, should be brought into play as early as possible, aiming for quick extinguishment and making full use of modern pumping capacities. But no matter how large or powerful a heavy stream may be, it will not serve its purpose unless it can do two things: reach out far enough to put water on the fire effectively, and apply enough water to the burning material so that heat will be absorbed faster than it is produced.

Many things determine the reach of a stream. The two most important limiting factors are defective nozzles and improper pressure at the nozzle. Either one of these will cause the stream to break up into fine particles, losing its force and solidity and simply dissipating into the air before reaching its target. Another common mistake is trying to reach too high from ground level. Based on many tests, we know that the maximum effective reach for all streams, including hand lines, comes when the stream travels at an angle between 30 and 35 degrees above horizontal; an elevation of 50 degrees marks the top limit for any real use of a stream. As a rule of thumb, we say that any stream directed from street level will not be effective above the third story of a building. Even here, reach may be sharply reduced by obstructions, such as interior walls.

The ability of a stream to get close enough to the burning material to absorb heat at an effective rate is called penetration. To be effective, a solid stream must put 90 percent of its water within an imaginary 38-cm (15-inch) circle drawn around the seat of the fire. Generally, the stream with the greatest striking force will have the greatest penetration. This is why heavy streams, when properly directed, have so much more effect on fires than smaller and less powerful streams. Automatic and remote controls for heavy streams have greatly improved our ability to direct and control them. Bear in mind, however, that once a heavy stream has served its purpose, it should be shut off at once both to conserve water and to prevent excessive water damage to the structure and its contents.

Hand Line Streams

Hand line streams can be either fog or solid. Practically everything just said about the factors affecting the reach and penetration of heavy streams applies to hand lines as well. But mobility, choice of nozzle, nozzle pressures, hose sizes, relationship of nozzle size to hose size and length of hose line, and procedures for stretching hand lines are important elements in using this fire fighting equipment.

The biggest nozzle that should be used on a hand line is 1¼ inches. The smallest sizes for satisfactory use are the ³⁄₁₆- and ⅜-inch nozzles used at the end of booster pump lines. Some nozzles not only allow the operator to shut off the flow of water but can also be shifted to deliver either a solid stream or fog; certain kinds can also be used with applicators of different types. (See Figure 6-1.)

There is also a definite relationship between nozzle size, hose size, and length of line that must be observed for effective use. A "short" line of hose contains up to six standard 15 m (50-foot) lengths of hose; six to twelve lengths is a "medium" line and more than twelve lengths

a. Solid stream triple stacked nozzle tips.

b. Solid stream nozzle with shutoff.

c. Solid stream nozzle-playpipe combination (with shutoff).

d. Bresnan circulating cellar nozzle.

FIGURE 6-1. Solid stream and combination. Solid stream and fog nozzles. Photographs Courtesy of the Elkhart Brass Manufacturing Company Inc.

180

CHAPTER 6

e. Combination fog-solid stream, constant flow nozzle with variable gallonage control.

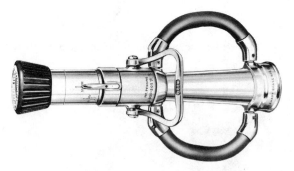

f. Remote controlled combination fog-solid stream nozzle.

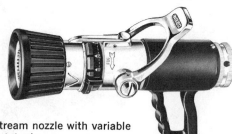

g. Combination fog-solid stream nozzle with variable gallonage control and pistol grip.

h. Combination fog-solid stream nozzle with playpipe. The tip has three calibrated straight stream positions, $\frac{7}{8}$", 1" and $1\frac{1}{8}$", which at 100 psi produces 175, 225, and 250 gpm, respectively.

FIGURE 6-1. Continued

is considered a "long" line. Passing over questions of hydraulics and pumper pressure, there are definite limitations on the size nozzle that can be used with any given length of line. For a short line, the largest nozzle size is one-half the diameter of the hose in use or, in figures, a 1¼-inch nozzle on a 2½-inch hose. For medium lines, the largest-size nozzle should be the next size smaller than one-half the hose diameter. This would mean that for a 2½-inch hose, the nozzle should be 1⅛ inches or, for a 3-inch hose, 1⅜ inches. A long line of 2½-inch hose, however, can take no more than a 1-inch nozzle. These guidelines apply to plain nozzles only, not to nozzles for fog or foam lines or the fixed nozzles attached to booster pump lines, as they are predicated on the volume of water passing through the hose line.

Fog

Using water in the form of fog, or spray, gives us the advantage of greater and quicker absorption of heat for every gallon used than we can get when we use a solid stream, because a greater surface is exposed. Fog has its greatest effect when it is applied cold and the heat of the fire evaporates it into steam. One pound (.45 kg) of water has a potential cooling effect of 1184.0 kjoules [1122.3 Btu (British thermal units)], the result of the combination of water's *specific heat* and *latent heat*. We mentioned these two types of heat at the start of this chapter; let us now consider them in some detail.

Water comes from the mains and out of a nozzle at about 15.6°C. (60°F.). Its *specific heat* is measured by the number of Btu required to raise one pound of water one degree F., at sea level. To climb from 15.6°C. (60°F.) to 100°C. (212°F.), at sea level, takes 353 kJ/kg (152 Btu) of heat. Water's *latent heat* is also measured in kilojoules per kilogram or Btu, but latent heat is the quantity of heat given off or absorbed in passing from the liquid state, water, to the vapor state, steam, or vice versa. This change amounts to 2257 kJ/kg (970.3 Btu). Together, the specific and latent heat of a pound of water add up to a heat absorption of 2610.2 kJ/kg (1122.3 Btu). Notice that it is this high latent heat of water that accounts for more of this total, an amount substantially greater than can be found in most common substances. This is what makes water such an efficient extinguishing agent.

Since fog vaporizes more rapidly than a solid stream of water, its high rate of heat absorption makes it a most effective extinguishing agent. Fog, properly applied, rapidly cuts off the heat side of the fire triangle (or tetrahedron); it absorbs heat that might otherwise extend combustion. Some authorities argue that fog substantially decreases the supply of atmospheric oxygen available to support combustion, another side of the fire triangle (see Chapter 1 for discussion) because water expands 1,700 times when vaporized. The resulting pressure, developing within a structure, will force heat and combustible vapors or gases out of the building.

Fog has further advantages:

- Reduction of smoke concentration
- Absorption, dilution, or chemical change of some toxic gases
- Protection of fire fighters advancing behind it
- Requires only one person to handle the nozzle
- Preserves evidence of arson
- Cuts down on water damage to structure and contents
- Conserves water, particularly for booster equipment
- Reduces conductivity characteristics

Fog is also particularly useful, when used with applicators that have angled heads, in getting at fire in enclosed spaces, such as ceilings, walls, and lofts. Figure 6-2 shows several types of fog applicators. Fire fighters need to make openings only large enough to admit the applicator. The fog, as it vaporizes into steam, tends to follow the fire along hidden channels. Used against fire in the chimney of large oil-burning heating systems such as are found in tall buildings, a fog applicator, extended through clean-out or other access doors near ground level, saves under-taking the dangerous operation of trying to force water down a high chimney from the top. Low velocity fog, generated by decreasing pressure in the hose line and increasing the spray angle of an adjustable nozzle, is favored for use against fire in places that use or store radioactive materials because it does not wash these dangerous substances about.

Fog, or spray, streams are also useful in protecting external exposures with a water curtain. Used in a confined fire area, a heavy fog application, the so-called "indirect method of attack," is a good procedure. Fog streams also find good use against fire in a building that is threatening to collapse, since they can be applied from outside the structure at considerably less risk to the fire fighter than if a hand line were taken inside.

In spite of these many advantages and the claims advanced in its favor by many experts in fire science, fog is not the ultimate or only effective way to use water as an extinguishing agent. Fog has certain disadvantages: it cannot be aimed as well as a solid stream, wind and gas currents can divert it away from the seat of a fire, and it tends to escape through any vertical opening in a structure by being carried upward by rising currents of heated air or hot gases. Fog also does not have as good a vertical or horizontal range as a solid stream; in other words, it will not reach up as high or out as far.

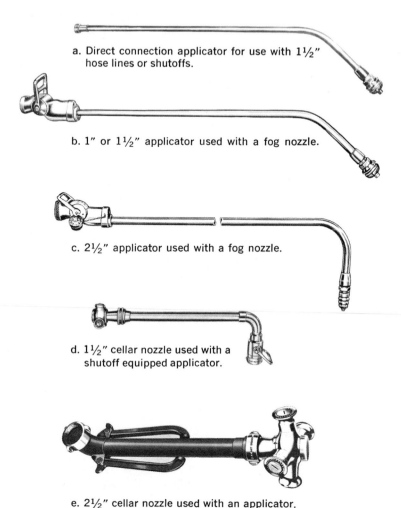

a. Direct connection applicator for use with 1½" hose lines or shutoffs.

b. 1" or 1½" applicator used with a fog nozzle.

c. 2½" applicator used with a fog nozzle.

d. 1½" cellar nozzle used with a shutoff equipped applicator.

e. 2½" cellar nozzle used with an applicator. The applicator has folding arms.

FIGURE 6-2. Several types of fog applicators. Photos a, b, and c courtesy of the Akron Brass Company. Photos d and e courtesy of the Elkhart Brass Manufacturing Company, Inc.

Foam

Fire fighters use two types of foam: chemical and mechanical. Chemical foam is formed by the reaction between sulfuric acid and sodium bicarbonate plus a stabilizing agent; together, this produces tough bubbles filled with carbon dioxide, an extinguishing agent in its own right. The chemicals that enter into the foam come as a dry powder, packed in cans. The contents of these cans are poured into the hopper of a foam generator that has water passing through it. The chemical reaction begins as soon as the powder gets wet. The amount of powder to water is proportioned in the generator, but the foam is mixed and expanded in the hose line between the generator and the nozzle. The foam, in turn, is delivered from the nozzle onto the fire.

Mechanical foam, or air foam, is generated by proportioning a foaming agent and a water stream with a strong beating action. The resulting foam is similar to what we see in a dishpan when we use a liquid detergent.

High-expansion foam is generated by a machine that blows large quantities of air through a fine mesh that has a soapy mixture flowing over its surface. The process is similar to blowing bubbles with a toy bubble pipe.

Actually, whether the bubbles of foam contain inert gas or air has little effect on the ability of foam to extinguish a fire. Foam's main value is in fighting fires in flammable liquids. It acts in several ways: blanketing the surface to cut off the atmospheric oxygen supply to the fire, insulating the liquid from the heat of the fire, and cooling its surface to reduce the liquid's production of flammable vapors.

Foam is effective against hydrocarbons that are liquids at ordinary temperatures and pressures, such as gasoline, kerosene and oil, but not against liquefied gases. Also, fires involving alcohols, ketones and the more volatile esters call for the use of special foams because their solvent action destroys ordinary foam. To be effective, foam must have a lower specific gravity than the liquid it blankets, should spread or flow easily across liquid surfaces (low viscosity), and must not break down rapidly when exposed to fire or on contact with the flammable liquid.

Knowing how to mix and apply each type of foam will largely determine whether the foam holds up or deteriorates rapidly. Bear in mind, however, that different types of foam must never be mixed. Mixing types of foam can damage equipment, may result in total loss of capability to generate any foam at all, and will certainly present a serious clean-up and maintenance problem.

Production of foam depends on using one of three foam-liquid concentrates: low-expansion, high-expansion, and alcohol-resistant. Regular foam concentrate is a dark brown liquid produced by chemical processing of protein materials, such as horn and hoof meal or chicken feather meal. Additives lower viscosity, prevent freezing at ordinary temperatures, prevent the foam from breaking down, and increase its resistance to fire. Foam concentrate is added to water through a flow proportioning device that may be relatively simple or quite complex.

One type of proportioner can be built into the pumper itself. Another type connects to the pumper outlet, and still another is an in-line eductor in the hose line. (See Figure 6-3.) The simplest and easiest type to use works in much the same way as the attachment for insecticides and sprays that goes on a common garden hose. Figure 6-4 illustrates this system; foam nozzle with a pickup tube draws foam concentrate from a 19-liter (5.0-gallon) can. Every 606 liters (160 gallons) of water will produce about 6245 liters (1650 gallons) of foam.

Regardless of type, all of these proportioners will automatically gauge the correct amount of foam concentrate, three or six percent, that is to be added to the water. The result is low-expansion foam, so called because the expansion rate is up to 10-to-1. This type of foam is commonly used to protect tank farms and to combat aircraft crash fires and most hydrocarbon fires. Where water is scarce, high-expansion concentrates will give between 16-to-1 and 20-to-1 expansion, with a high rate of flow and coverage. This type of foam is particularly effective against spill fires.

Where a hand line is big enough to fight a flammable liquid fire, there are special play-pipes that can deliver a straight stream, or fog, or both, at the operator's discretion. Effective range varies from 6.1 to 48.8 m (20 to 160 feet); maximum flow, from 57 to 3785 lpm (15 to 1000 gpm). Used with a foam concentrate attachment, this is an ideal foam generating system for use with apparatus that has a booster pump and booster hose. It can also be used on any layout of hose that can be made up to carry a 1½-inch fitting and still give the required pressure at the pickup tube.

How much foam and how fast to apply it depends on the size of the fire, the characteristics of the flammable liquid and the method of application. A small tank may need only a few inches of foam to cover it effectively; a large tank may mean blanketing with several feet of foam. Hand lines are limited in their use against large flammable liquid fires. Tank farms, where flammable liquids are stored in bulk, often have foam makers mounted near the top of each oil-storage tank. In case of fire, foam solution can be supplied from a safe distance through pipelines. In a pinch, fire fighters may be able to approach behind a shield of water to where they can substitute hose for damaged or broken pipelines. Tanks may also be protected by portable foam towers. These devices have a long riser pipe that ends in a gooseneck outlet that can be hung over the top of the burning tank. Portable towers are not as good a method of giving tanks foam protection because the intense heat an oil fire generates may prevent moving a portable tower close enough to be of any use.

True high-expansion foams are foams with an expansion ratio exceeding 100-to-1. Actually, 100-to-1 is a low expansion, compared with the

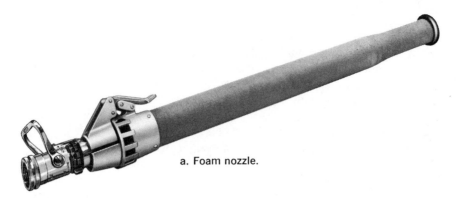

a. Foam nozzle.

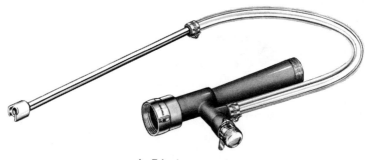

b. Eductor.

c. Eductor with foam bypass.

FIGURE 6-3. Foam nozzle and in-line educators. Courtesy of the Elkhart Brass Manufacturing Company, Inc.

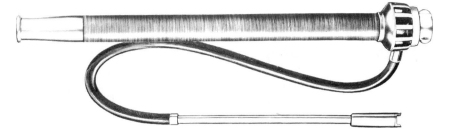

FIGURE 6-4. Foam nozzle with a pickup tube. Courtesy of the Akron Brass Company

foams used by some fire departments that have a 1000-to-1 expansion ratio.

High-expansion foam has a number of uses. Its use as a fire extinguishing agent can best be summed up by a statement which appeared in the January, 1969, Fire Protection Equipment List of the Underwriters' Laboratories, Inc.:

> High-expansion foam may be usable in extinguishing fires in ordinary hydrocarbon fuels and in controlling fires in ordinary Class A combustibles when (1) the hazard is moderate, (2) foam covers the hazard in a maximum of 10 minutes, and (3) foam is maintained over the hazard for not less than one hour. The 10 minute fill time may need to be increased to a longer time, depending on the construction of the building and the severity of the hazard.

Many smaller and most large fire departments now have high-expansion foam units. Some departments refer to these units as "bubble machines" because the foam produced looks like soap bubbles.

Fire departments are equipped for both local and total flooding application of the foam. Water is generally fed into the high-expansion foam units through ordinary fire hose. A proportioner draws the foam concentrate from containers, meters the amount required, and feeds the concentrate into the water stream. The foam is discharged through large tubes, covering the fire; flanged tubes are available for doorway application.

On application, the foam will both smother the fire and cool the burning material. The water in the foam is converted to steam which absorbs the heat and aids in reducing the oxygen content of the surrounding air.

Applying high-expansion foam in the open with a strong wind blowing has proved to be difficult. This difficulty has been partly overcome by lowering the ratio of foam to water. This change builds up foam resistance to the tearing action of the wind.

High-expanison foam has proved effective in protecting industrial complexes. It has been used successfully as a method of controlling fires and suspected fires in dust collecting systems. A good example of a total floor industrial protection system is a warehouse, 213 meters (700 feet) long, containing 42, 475 cubic meters (1,500,000 cubic feet) of space, in Jacksonville, Illinois. This warehouse is protected by a total flooding system, capable of completely filling the building with foam in seven minutes. This is a good example of private fire protection (see Chapter 8) with a high-expansion foam system.

Although high-expansion foam has proved useful, it has limitations: In addition to being affected by high winds, it is adversely affected by fire wind currents and fire intensity. It should be considered as another tool available for use in a particular situation.

"Light water," newest of the fire extinguishing agents, has been referred to as the most significant advance in fire suppression in recent years. The term "light water" is a trade name of the 3M Company.

"Light water" is a noncorrosive, nontoxic fluorochemical liquid foaming concentrate that mixes with water and can be aerated into foam. Recommended mixing proportion, by volume, is 6 percent concentrate to 94 percent water. The agent-water solution has an expansion rate up to 10-to-1. The foaming agent makes water float on gasoline and jet fuel, hence the name "light water." Its primary use is in extinguishing fire, although it can be used effectively to prevent ignition or reflash in flammable liquids.

Development of "light water" began late in 1962 at the Combustion Suppression Research Center of the Naval Research Laboratory in Washington, D.C. It was developed primarily for aircraft crash fire fighting but undoubtedly will be adopted for many other uses. The material reached naval field stations and was put into use late in 1965.

This new type of foam has been exhaustively field-tested by the U.S. Naval Research Laboratory, the U.S. Naval Air Systems Command, the Los Angeles Fire Department, the Canadian Armed Forces, and the Ministry of Technology in the United Kingdom. All tests have demonstrated that the material is far superior to protein foam. The following is an example of a test conducted by the U.S. Navy:

A 21-meter (70-foot) diameter fire, fueled with 4542 liters (1200 gallons) of aviation gasoline, was allowed a preburn time of 43 seconds. This provided maximum heat buildup. Within 11 seconds after fire fighting operations began, the fire was sufficiently under control to permit rescue operations and was completely extinguished in 53 seconds. During the entire test, only 45 liters (12 gallons) of 6 percent concentrate "light water," proportioned with 700 liters (185 gallons) of fresh water, were used. Protein foam concentrate was then used to attack and extinguish a fire developed under identical conditions. By contrast, it took two complete truckloads, each holding 189 liters (50 gallons) of protein foam concentrate, proportioned with 3030 liters (800 gallons) of water, to bring the fire under control in 90 seconds and to extinguish it in 172 seconds.

Results of tests conducted by the Los Angeles Fire Department were similar. These tests showed "light water" to be 2.2 times more effective than protein foam. Effectiveness was measured by the time required to extinguish fires of the same nature and intensity in two different tests. The 2.2 figure was an average for all fires. Many of the officers observing the tests commented that "light water" appeared to have the quick knockdown characteristics of Purple K and the holding characteristics of protein foam.

"Light water" has primarily been used for aircraft crash fire fighting although experiments have been conducted as to its value in extinguishing tank fires. Conservatively, test results have proven it to be 2 to 3 times more effective than protein foam. It has been tested and shown to be compatible with Purple K. The U.S. Navy is planning to use it for shipboard fire fighting, particularly aboard carriers. The Aviation Committee of the NFPA recommended to the NFPA membership, at the 1969 Annual Meeting, that Aqueous-Film-Forming-Foam (AFFF) (light water) be recognized on an equal basis with conventional foams as a primary extinguishing agent for aircraft rescue and fire fighting purposes.

Foam is useless against fires that involve oxidizing agents and should not be used on combustible metal fires or on energized ("live") electrical equipment.

Wetting Agents

Wetting agents are solutions that, when added to hard or soft water or to salt or fresh water, reduce surface tension by almost two-thirds. The result of this decrease is that the penetrating quality of the water increases greatly; less needs to be used. Added to fog, a wetting agent speeds the change from droplets of water to steam, creating even quicker absorption of heat and faster extinguishment of the fire.

Wetting agents make structural overhaul go faster and are particularly effective in preventing rekindles. Overhaul of contents is made simpler and faster because such things as upholstered furniture and baled materials do not have to be opened up as much as when only plain water is used to extinguish fire. At a fire where heavy smoke is present, wetting agents, particularly when used with fog, reduce life hazard by increasing the absorption of smoke particles that are suspended in the atmosphere.

The most serious disadvantage of wetting agents is that they tend to destroy or lessen the effect of salvage or protective covers. Although wetting agents that meet NFPA standards are no more corrosive than plain water to steel, brass, bronze, and copper, some types are corrosive; others are detergents that accelerate corrosion although they do not cause it themselves. Some wetting agents tend to reduce the range of solid streams and to affect both the range and pattern of fog streams. Again, those meeting NFPA standards do not seem to cause any appreciable change in the range of solid streams or fog, nor do they reduce the angle of discharge for fog nozzles by more than 10 percent.

Dry Chemical

Dry chemical is a powder mixture that is used as an extinguishing agent, and may be applied through portable extinguishers, hand-held hose lines, or fixed pipe systems. The chemical mixtures are basically free-flowing powders and may consist of sodium bicarbonate, potassium bicarbonate, potassium chloride, urea-potassium bicarbonate, and mono-ammonium phosphate. Various additives may be added to these mixtures to improve their storage, their flowing, or their water repellency characteristics.

The terms "regular dry chemical" and "ordinary dry chemical" generally refer to powders that are listed for use on Class B and Class C type fires. "Multipurpose dry chemical" refers to powders that are listed for use on Class A, B, or C fires. The terms "regular dry chemical," "ordinary dry chemical" and "multipurpose dry chemical" should not be confused with "dry powder" or "dry compound," which refer to those powders designed for use on combustible metal, or Class D, fires.

A more complete description of dry chemicals can be found in the NFPA *Handbook of Fire Protection,* 14th edition, and by reading *Fire Suppression and Detection Systems,* by John L. Bryan, published by Glencoe Press (Figure 6-5).

Since these powders will flow almost like water, they can be stored in extinguishers or tanks until they are needed. An inert gas, such as nitrogen or CO_2, stored under pressure, is used as the propellant to apply the powder to a fire. Extinguishers come in sizes from 1.1- to 1814-kg (2½- to 4000-pound) capacity; some have self-contained propellant cylinders, others depend upon attached cylinders or separate tanks that are connected to the powder container by rigid or flexible piping. There are even chemical trucks that carry several tons of powder and use turret nozzles to put out 11 kg (25 lb) of powder per second over distances up to 27 meters (90 feet).

Dry chemical extinguishers can be used against surface flames of fires in most combustible materials but their best use is against flammable liquid fires and liquid petroleum gas fires. Although dry chemicals will not extinguish a fire permanently unless they can cover the entire surface almost simultaneously, this frequently happens. The powder comes from the nozzle in large clouds, spreads over a wide area very rapidly,

Class "A" fires occur in ordinary combustible materials such as wood, cloth and paper. The most commonly used extinguishing agent is water which cools and quenches. Fires in these materials are also extinguished by special dry chemicals for use in Class A, B & C fires. These provide a rapid knock down of flame and form a fire retardant coating which prevents reflash.

Class "B" fires occur in the vapor-air mixture over the surface of flammable liquids such as greases, gasoline and lubricating oils. A smothering or combustion inhibiting effect is necessary to extinguish Class "B" fires. Dry chemical, foam, vaporizing liquids, carbon dioxide and water fog all can be used as extinguishing agents depending on the circumstances of the fire.

Class "C" fires occur in electrical equipment where non-conducting extinguishing agents must be used. Dry chemicals, carbon dioxide, and vaporizing liquids are suitable. Because foam, water (except as a spray), and water-type extinguishing agents conduct electricity, their use can kill or injure the person operating the extinguisher, and severe damage to electrical equipment can result.

Class "D" fires occur in combustible metals such as magnesium, titanium, zirconium and sodium. Specialized techniques, extinguishing agents and extinguishing equipment have been developed to control and extinguish fires of this type. Normal extinguishing agents generally should not be used on metal fires as there is danger in most cases of increasing the intensity of the fire because of a chemical reaction between some extinguishing agents and the burning metal.

FIGURE 6-5. Classification of fires.

and leaves a glassy coating on solid materials that effectively prevents rekindling.

For a long time, even experts thought that the success of dry chemicals depended on a sort of "sweeping action" that literally brushed the flame from the surface of the burning material. Next came the theory that fine particles produced a special high cooling effect or that the chemical reactions involved produced enough inert gas, principally carbon dioxide, to cut off the supply of oxygen to the flame. Fire science chemists objected to all these theories because a given amount of dry chemical would put out more fire than it theoretically should.

After long and careful investigation proving that the success of dry chemical extinguishers did not depend on the chemical's ability to break any one of the three sides of the conventional fire triangle, fire science research discovered the answer. During combustion, molecules break up into simpler fragments called "radicals." When these fragments combine with atmospheric oxygen, they transfer additional energy to neighboring molecules, continuing and building up a reaction-chain that helps the combustion process along. What the dry chemical did, research discovered, was to stop this transfer of energy and react with the radicals. This also explained why Purple K was more effective, pound for pound, than sodium bicarbonate: potassium is a more active element than sodium.

Recent research has discovered a new technique for using dry chemical extinguishers. Faced with fire in large spills of flammable liquids, those fighting the blaze should apply dry chemical first, followed immediately by foam. The results are almost too good to be true. By using foam, the quick extinguishment of the surface flame by the dry chemical can be made permanent. The only drawback to this two-pronged attack is that the combination is limited to certain dry chemicals only. These are called "compatible" powders and do not attack the foam bubbles.

The misuse of various powder mixtures on fires has been known to increase the fire and cause personal injury. The mixing together of one chemical with another has also contributed to severe personal injury and should not be done without full knowledge of the compatibility of such mixtures.

Carbon Dioxide (CO₂)

Carbon dioxide (CO_2) is an inert gas. It will neither burn nor support combustion. Since it is heavier than air, it will settle close to the ground and cut off the supply of oxygen any fire needs to keep going. As an extinguishing agent, it has many advantages: it is cheap to make, leaves no residue, is nontoxic, a nonconductor of electricity, noncorrosive, and is easily compressed into small containers. CO_2 extinguishers find their best use against Class C fires where a live electric current is present. But carbon dioxide should not be used against Class D fires, those involving combustible metals. (Figure 6-6.)

Carbon dioxide extinguishers contain the gas in liquid form. This means storage under pressures from 5510 to 6890 kPa (800 to 1000 psi). Sizes vary, but portable units can be large enough to carry 100 gallons of liquid gas. Larger sizes are often mounted on two-wheeled carts for ease in handling and use.

Regardless of size, the effectiveness of a CO_2 extinguisher depends on the ability of the gas to cut off the supply of oxygen to a fire. Even though the snow that appears when a CO_2 extinguisher is discharged has a temperature of –78.9°C. (–110°F.), it has a latent heat of 573.1 kJ/kg (246.4 Btu per pound). As only about 25 percent of the gas is converted to snow, the cooling effect totals about 279 kJ/kg (120 Btu per pound). Compare this effect to that of water, assuming all the water were converted to steam. This extends to the large mobile units, using refrigerated, low-pressure tanks, that can carry more than five tons of carbon dioxide. Fixed systems in plants and other buildings are used in preference to sprinkler systems when water could cause serious damage to goods, business records, or delicate instruments, or react with burning materials. (For detailed discussion of fixed systems, see Chapter 8.)

FIGURE 6-6. Examples of CO² and "snow." Courtesy of the Palo Alto Fire Department. Photo by Jim McGee

Halogenated Hydrocarbons

Halogenated hydrocarbons are extinguishing agents composed of carbon, hydrogen, and one of several poisonous gases, fluorine, chlorine, bromine, or iodine. They are applied to a fire by pumping or gas pressure and the heat of combustion makes them boil or vaporize rapidly. Their effectiveness depends on their ability to interrupt the chain of combustion. In vapor form, these chemicals change to other chemical compounds that inhibit or prevent oxidation. Because these compounds are nonconductors, they are particularly useful against Class C electrical fires; they also do well against some Class B fires.

Three of these agents have been in use a long time: carbon tetrachloride, chlorobromomethane, and methyl bromide; others are more recent developments. Whatever their history, however, there are many serious objections to the use of these compounds as extinguishing agents. Several release highly toxic gases when heated; others emit toxic vapors at room temperature. A number of them attack and very badly corrode metals after only a short exposure and, if water is also used or present, the reaction becomes even stronger. Almost every one of these extinguishing agents will cause skin and eye irritation or damage the respiratory tissues. Some can cause serious tissue damage, much like a burn. Because of the seriousness of the toxicological hazards and because other safer extinguishing agents became available, these and other similar hazardous agents have been eliminated from the NFPA, UL, and ULC listings for extinguishing agents. However, carbon tetrachloride may still be found in older, obsolete extinguishers and sold as electrical/electronic equipment cleaning agents because of its nonconductive characteristics.

Halon 1301 (Bromotrifluoromethane, $BrCF_3$), the least toxic and one of the most effective extinguishing agents, has been developed for use in portable fire extinguishers. However, unless a very small specific fire area is to be protected, the replacement costs normally preclude its vast commercial usage. The low toxicity of this agent permits the rapid and automatic discharge of the agent into occupied spaces, prior to evacuation and at the first detection of a fire, making this particular agent very valuable. This agent is rated approximately 2½ times as effective as CO_2 used in similar installations.

Dry Powders

These extinguishing agents are our best answer to the problems raised by Class D fires, combustible metals. Although they are marketed under a great number of trade names, almost all dry powders depend on the same principles for their effect. We will limit our discussion to typical examples of the different types and how they work.

Pyrene G-1 contains powdered graphite and phosphorous compounds. When heated, the phosphorous compounds generate vapors that cut off the oxygen supply to the burning metal; the graphite cools the metal itself to below its ignition point by conducting heat away. G-1 and the powders that resemble it have serious drawbacks: they can only be applied with a shovel or a handscoop, which means one must get very close to the burning metal, and, at the same time, the powder must be applied with care, particularly where the burning metal is in the form of turnings, chips or dusts, to prevent spreading or explosion. For best results, the dry powder should be built up to form a crust about an inch thick over the burning material and allowed to remain undisturbed until the metal cools down.

Another dry powder, Ansul Met-L-X, contains a sodium chloride or table salt base, with an additive to prevent caking, and a plastic that melts and fuses the powder into a solid cake over the burning metal. Met-L-X and similar dry powders come in an extinguisher but the range is only about 10 feet. Further, great care must be exercised not to blow bits of the burning metal around with the ejected powder. Met-L-X's effectiveness depends on cutting off the oxygen supply to the burning metal. It also has the great advantage of clinging to vertical surfaces, a very helpful property when used against a combustible metal fire involving large castings or pieces. Table 5 compares the effectiveness and limitations of these two types of dry powder against certain metal fires.

Although they are not dry powders, our discussion must include the relatively new liquid extinguishing agents, some still in the experimental stage, that are used against combustible metal fires. Perhaps the best of these agents, TMB (trimethoxyboroxine) liquid, developed for the U.S. Navy, has shown considerable effectiveness against magnesium fires and some value against zirconium and titanium fires. Applied as either a stream or spray from 9.5-liter (2½-gallon) pressurized extinguishers, TMB can reach into places dry powder cannot, such as engine compartments, and can be applied from safer distances.

TMB's effectiveness depends on a breakdown of its chemical compound that coats the burning metal with molten boric oxide. This coating prevents formation of combustible metal vapors at the surface of the burning material and also cuts off the oxygen supply to the flame.

TABLE 5. EFFECTIVENESS OF PYRENE G-1 AND ANSUL MET-L-X, COMPARED

TYPE OF FIRE	COMPLETE EXTINGUISHMENT	
	Met-L-X	G1
1. Magnesium chips or turnings, dry or oily	Yes	Yes
2. Magnesium castings or shaped pieces	Yes	If pieces are covered
3. Titanium turnings, dry or oily	Yes	Yes
4. Uranium solids and turnings	Yes	Yes
5. Zirconium, chips and turnings, when coated with water-soluble oil	Yes	Yes
6. Zirconium chips, moist	No, control only	No, control only
7. Sodium, in depth or spills	Yes	Yes
8. Sodium, spray or spill on vertical surface	Yes	Unsatisfactory
9. Potassium, or alloy of sodium and potassium spill	Yes	Yes
10. Same as (9) but in depth	Yes, but with difficulty	Yes, but with difficulty
11. Lithium spill	Yes	Yes
12. Same as (11) but in depth	Unsatisfactory	Yes
13. Aluminum powder	Yes	Yes

Surprisingly, TMB liquid is flammable. When applied to a combustible metal fire, a heat flash follows, caused by the breakdown of the chemical compound. Fire in the metal quickly dies, followed by a secondary, quite green methanol flame that lasts only a short time and is self-extinguishing.

Using TMB liquid offers the fire service a major advantage in fighting fire in combustible metals: as soon as flames from the combustible metal cannot be seen, we can use water to cool the heated mass. Low pressure fog is perhaps the best form in which to apply water because care must be taken to keep from breaking down the boroxide coating on the metal itself.

REVIEW QUESTIONS

1. What is the most important and widely used extinguishing agent and how does it put out a fire?
2. What are two things essential to obtaining extinguishment with a heavy stream?
3. Name two types of streams of water used with hand lines.
4. Name five advantages of the use of fog streams.
5. How are the following types of foam made and used in fire fighting and on what types of fire are they effective?

    (a) chemical foam

    (b) mechanical foam

    (c) high-expansion foam

6. What is light water and how is it used?
7. How are wetting agents used in fire fighting and under what conditions are they most effective?
8. On what types of fires are dry chemicals effective?
9. How is carbon dioxide used in fire fighting and what are its advantages?
10. What are halogenated hydrocarbons?
11. What are the disadvantages of the first halogenated agents?
12. On what type of fires are dry powder extinguishers used?

7 Fire Fighting Tactics and Strategy

When an alarm sounds, summoning crews and apparatus to the scene of a fire, it sets in motion a chain of events that continues until the fire is out and the fire fighters are back in quarters. Any company responding to a call may find a situation that will test its skills, ability, and courage to the utmost. Lives may be in danger, buildings and property may be threatened with complete destruction, and there is always the danger that any fire may spread.

To prevent loss of life and property, fire fighting operations must be efficient, carefully controlled, and well directed. Efficiency depends largely upon training and good apparatus; control and direction depend on a sound grasp of fire fighting tactics and strategy. Never forget: the choice of tactics by the "first-arriving" company officers may mean the difference between a fire that is quickly extinguished, with little loss of any kind, and a major fire. And the choice of correct tactics depends on preplanning.

Preplanning

As we saw in Chapter 4, the biggest part of any fire department's inspection program is carried out by the members and officers of individual companies. Besides being a major factor in fire prevention, these inspections serve to give officers and crews firsthand information about types of occupancy, building construction, contents, and exposures. When supplemented with maps and sketches of the area, reports of inspections (see Figure 7-1) form the basis of plans for fighting fire in any given location. This advance information, like intelligence reports on enemy installations, allows those in command to make rapid decisions once they arrive at the scene of a fire.

FIRE DEPARTMENT
PREPLANNING INSPECTION FORM

Street No. _____

	HOUSE KEEPING	INSPECTED	
		DATE	BY

ADDRESS _____ DATE _____

FIRM OR BLDG. _____

OWNER OF BLDG. _____ ADDRESS _____

BUSINESS MANAGER _____ HOME PHONE _____

RESPONSIBLES IN EMERGENCY

_____ HOME PHONE _____

_____ HOME PHONE _____

_____ HOME PHONE _____

OCCUPANCY _____ TYPE OF CONSTRUCTION _____

ROOF: TYPE _____ COVERING _____ FLOOR _____

HEIGHT _____ CEILING _____ INT. WALLS _____

SPRINKLERED _____ STANDPIPES _____

CONTROL BD. _____

FIRE DETECTION EQUIP: TYPE _____ MONITORED BY _____

HEATING SYSTEM: TYPE _____ CONTROLS LOCATION _____

AIR COND.: TYPE _____ CONTROLS LOCATION _____

ELECT. SUPPLY: VOLTS _____ CONTROLS LOCATION _____

ATTIC ACCESS _____ AREA _____ SPRINKLERED? _____

BSMT ACCESS _____ AREA _____ SPRINKLERED? _____

GENERAL STORAGE AREA _____

FLAMMABLE LIQUID/HAZARDOUS STORAGE _____

HAZARDOUS PROCESS AREAS _____

VERTICAL OPENINGS _____ PROTECTED? _____

EXTERIOR EXPOSURES _____

MISCELLANEOUS/REMARKS _____

FIGURE 7-1. Preplanning inspection form.

Basis information of preplanning includes: geography, weather conditions, building locations, traffic conditions, types of occupancies, types of structures, building layouts (floor plan and location of equipment or materials), built-in fire protection, and information about utilities.

Geography and weather

"Geography," as used in fire science, means the physical features of a company's district, such as hills, narrow bridges, railroad tracks, rivers, one-way streets, overhead obstacles, and so on. Studying the geography of a district helps to determine the best avenues of approach for apparatus, the fastest running time, alternate routes to avoid heavy traffic, and ways to allow for needed detours or other contingencies. In some climates, snow, rain, wind, and temperature conditions at each season of the year will affect preplanning, not only for individual company officers but also for any subsequent operations involving a large force. For example, in a hilly city where snow and ice can be expected during the winter season, good preplanning would include routes for apparatus that will avoid or reduce the chances of getting stuck or being unable to climb to the fire site.

Traffic conditions

Knowing probable traffic conditions around a fire building helps the company officer to place his apparatus rapidly, a most desirable thing. These conditions include street parking, pedestrian traffic, and the flow of passenger cars and trucks. If the fire is in a building where traffic normally clogs the streets, preplanning should contain provisions for alerting the police to divert and control vehicular traffic.

Building location

The location of a fire building, in relation to its surrounding exposures and the terrain, should be known before a fire occurs. Where a building stands on a steep hill or is down in a depression, positioning fire apparatus may be seriously handicapped or limited. Similarly, surrounding utilities, such as electric power lines or gas installations, may also hamper placing the apparatus. If these conditions are not known before an emergency occurs, the resulting delay can have serious results.

In addition, at least part of initial fire fighting tactics depends on the relative location of any given building to surrounding exposures. Where the nearby buildings are of frame construction, well within possible reach of flames, immediate attention must be given to preventing the spread of the fire to those highly flammable structures. A nasty complication of this problem arises when the building on fire is one with many occupants, such as a school or hospital. Exposures close at hand may seriously hamper getting the occupants out or prevent assembling them in a safe place where they can be counted, a guarantee that the building is clear.

On the other hand, when exposed buildings are of fire-resistive construction or have outside protection, preventing fire spread may not require more than limited effort and need not be undertaken at once. Remember that smoke and toxic vapors can expose surrounding areas too.

Building Construction

The type of building construction probably plays a greater part in determining fire fighting tactics than almost anything else. From inspection, the first-arriving company should know:

1. Type of construction
   (a) brick with steel supporting beams
   (b) brick and wood
   (c) all frame construction
   (d) reinforced concrete
   (e) other (possible combinations are almost infinite)
2. Number of stories and height of building in feet
3. Type of roof
   (a) flat
   (b) peaked
   (c) shed
4. Type of roof covering
   (a) asbestos shingle
   (b) tar and gravel
   (c) tar paper
   (d) metal
   (e) slate
   (f) wood shingle
   (g) other
5. Type of roof supports
   (a) wooden beams
   (b) steel beams
   (c) other metal beams, including open truss
   (d) prestressed concrete
   (e) other

6. Type of windows
   (a) steel casement
   (b) double hung
   (c) wooden frames
   (d) fixed glass (if fixed-glass panels or windowless buildings, are there panels for fire department access and can they be found readily?)
7. If building is air-conditioned or ventilated, where are ducts located and how can the system be shut down?
8. Does the building have a fire alarm system?
9. Does the building have a sprinkler system?
10. Does the building have a standpipe system?
11. Locations of exits, stairways, elevator shafts, utility ducts; gas, electricity, and fuel oil shutoffs; refrigeration systems shutoff and type of refrigerant; location of cellar and basement drains or intakes.
12. Type of construction of immediate exposures

This partial list of construction features (for a more detailed list, and examples of forms, see Appendix E) offers immediate information that helps determine the initial attack on the fire and how to make the best use of apparatus and personnel.

Type of Occupancy

Careful preplanning includes knowing these things about the buildings in a given district and the people in them:

1. Is the building industrial, mercantile, commercial, residential, public, or a combination of occupancies under one roof?
2. Is the building institutional, such as a hospital, jail, school, or nursing home?
3. When is the building occupied?
   (a) by day
   (b) by night
   (c) round-the-clock
   (d) more at one season than another
   (e) peak periods of use during any 24 hours

And there are things to know about the contents of the building:

1. Are they considered fast- or slow-burning?

2. Are there explosive materials in use?

3. Are floor load limits likely to be exceeded at certain times?

4. Is there heavy machinery in use?

5. If there are hazardous chemicals or flammable liquids in the building, what quantities are normally used?

6. Are there water-reactive chemicals in use?

7. Are quantities of hazardous materials stored near or around the building, creating an immediate exposure problem?

8. Do exposure buildings contain quantities of combustible or hazardous materials?

When we use the terms *fire fighting strategy* and *fire fighting tactics,* what do we mean? These terms are often misunderstood.

The term *fire fighting strategy* is used to indicate an overall plan to attack the fire, rescue occupants, and extinguish the fire. Three basic features to a strategic plan are: (1) what is to be done? (2) what is needed to do it? and (3) who is to do it?* When one adds another question to this plan, (4) how is it to be done, the plan now becomes a tactical plan.

Therefore, the term *fire fighting tactics* means "the art of directing and employing personnel, apparatus, equipment and extinguishing agents in the fire ground."† For the purpose of this text, we can simply say that the tactics used will be part of the overall strategy employed by the officer in charge. Together, tactics and strategy dictate how we use the personnel and equipment on the fire ground.

Various departmental standard operating procedures are often of strategic value, and in turn influence the choice of tactics. Adequate prefire planning is what gives the company officer the facts needed to make a wise choice of fire fighting tactics.

The choice itself may be based on personal experience, knowing what worked well in a similar situation; but the student of fire science will have an advantage since he or she can also draw on the experiences of others contained in case histories of past fires.

The point is that the first-arriving officer is technically in charge of operations until relieved by a superior officer. To conduct operations, this officer must have a plan and, to arrive at a plan, must make a decision based on facts. The officer must quickly survey and analyze

*William E. Clark, *Fire Fighting/Principles and Practices* (New York: Dun-Donnelley Publishing Corp., 1974).

†Lloyd Layman, *Fire Fighting Tactics* (Boston: NFPA, 1972).

the situation, weigh the different factors involved, apply basic principles, decide on what action the crews must take, and direct their efforts.

Size-Up

There is a strong resemblance between military and fire fighting operations. The commander of an infantry unit, for example, although having intelligence reports about the enemy position and armament, will make a careful reconnaissance before deciding on the proper tactics. The facts gathered on the reconnaissance form the basis of an estimate of the situation, updating and modifying the intelligence reports. Inspection reports serve the same purpose in the fire service as intelligence reports do in the military; preplanning partially replaces reconnaissance. To bridge the information gap that remains, before formulating a plan of action, the fire company officer goes through a process known as "size-up." This process begins the moment the alarm sounds.

Analysis of a good size-up reveals five parts:

- Facts
- Probabilities
- Own situation
- Decision
- Plan of operation

The analysis of a fire situation must be done quickly by the first-arriving fire officer. Basically, the analysis leads to certain choices, to possible decisions, to action and finally to the evaluation of a fire situation. The method used by this text (Lloyd Layman's method) is simple and easy to follow. This is not the only method in use today; however, all methods cover the same basic components.

The importance of a correct size-up cannot be overemphasized. The entire course of fighting a fire is largely determined by the decisions of the first-arriving company officer. Everything that follows is usually part of plans worked out before the fire occurred and those plans can be carried out only if the initial response decisions were correct. Size-up is so important that we will consider it in some detail, although the subject is covered fully in texts on fire fighting tactics and strategy.

Suppose you are looking over the shoulder of an officer whose company has a first-response assignment, just after an alarm comes in. If the alarm is phoned in and the caller can be kept on the line long enough to get full particulars, or if the alarm box is one that is tied directly into the fire department's alarm system, such as in a school or a major business building, our officer will know exactly what building is involved. More often, the alarm code will only serve to identify a particular block, or larger area. Locating the fire building may have to wait until the first-due company gets close to the scene.

Regardless of where exactly the alarm places the fire, our officer takes a quick look at the clock. This will tell whether danger to life is high, moderate, or low. During early morning hours, life hazard is highest in hotels, hospitals, and similar places where people are sleeping; during daytime hours, the number of people in mercantile buildings, industrial plants, and schools will be at a peak. Remember, though, that some plants operate round-the-clock, with full shifts. Knowing the time also tells what to expect in the way of traffic, whether to call for police help in clearing the streets, and whether an alternate route will be the best way to cut running time. The officer learns the facts about wind and weather, if they are not already known, when the apparatus rolls out of the station.

The season of the year is also a factor in size-up, because production of goods may reach a peak before holidays, stores may have extra stock on hand, temporary help may increase the normal number of people working in a given building, and neglected safe practices, particularly for trash disposal, may seriously increase fire hazards.

Seated beside the driver, our officer consults the company water resources map. Some people, after only a few months in a given protection district, can carry this information in their heads; most cannot. It is generally better to check the maps, rather than trust to memory. These maps show the hydrants or water resources in the area. Once the fire building is located, departmental standard operating procedures will pretty well govern where to place the apparatus. Our officer, if lucky, will have learned the precise location of the fire building via the radio during travel. This gives a little extra time to check the preplanning file. Normally, it will not be known until reaching the fire ground just what building is involved.

Reaching the fire scene, our officer sees new facts: the building and its surroundings, the amount of smoke, its color, where it is coming from, visible flame, what part of the building seems involved, and the presence or absence of people at the windows, on fire escapes and roofs, or in the street. And most important, the officer learns where the fire is burning inside the building.

Notice that both the location of the fire building and the point where the fire is burning inside are steps in size-up that cannot be preplanned. Furthermore, locating the fire is commonly considered one of the three basic elements of fire fighting strategy. This makes sense. Since we cannot begin to fight an enemy intelligently unless we know where the enemy is, location has a great bearing on our choice of tactics.

Coming back to our company officer: From what he or she sees, backed by fire fighting experience, the officer fixes the location of the fire on the first floor, at the rear of the building. Preplanning and familiarity with the territory now come into play as the size-up continues. The hazards posed by type of construction, age, and structural features of a building are generally the same. Old buildings, with open stairways and shafts, favor rapid spread of fire. Similarly, extensive use of wood for floor joists, flooring, interior walls, decoration, plus the absence of fire-stops between floors, makes for fast fire spread and structural weakness. Early loft buildings or old apartments converted to factory use are particularly dangerous. Figure 7-2 shows a typical early loft building.

In short, every building has its own characteristics and these features must be taken into account in planning the attack. Preplanning helps in making the size-up by ensuring that all construction features are taken into account before a decision is made. It also points up the presence and nature of any special hazards.

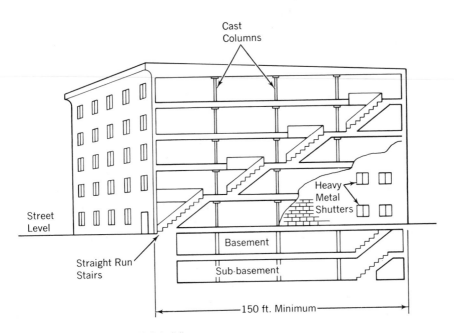

FIGURE 7-2. Typical old loft building.

After looking over the fire scene and taking the hazards into account, our officer estimates what the fire can do. From the type of construction, it can be known if the flames will spread rapidly. From preplanning, the officer knows whether the surrounding buildings present an exposure hazard, if there is access to the fire building, and whether to protect the exposures. The officer also knows if there are special hazards, such as potentially explosive materials, dangerous industrial processes, quantities of combustible materials, flammable liquids, or other similar threats to the safety of occupants or fire fighters; whether there are people working or living in the building, if so how many, and if they will need to be rescued; also, whether stairways may be unprotected and fire escapes inadequate.

The next question our officer faces is what resources are available to handle the situation. The answer to this question plays a large part in the decision of what action will be taken. One part of the answer is the kinds of companies responding, the number and types of apparatus, and the number of personnel available. How much water will be needed to extinguish the fire can be estimated by calculating how much steam would be required to expel all oxygen from the fire area. This can be estimated by determining the volume of the fire area and applying the known factor for conversion of water into steam. This yield formula, expressed as gallons per minute, is equal to the volume of the area divided by 100, and is very effective during this portion of size-up. This same basic formula can be used to determine the number of people needed to deliver the number of gallons of water required to be delivered. This field formula is expressed as the number of gallons of water per minute divided by 50 (the approximate number of gallons per minute handled by each person) and is equal to the number of fire fighters. These formulas and supportive documentation can be found in other texts.* Adequacy of water supplies must also be considered. Finally, the officer must know if help has been summoned, or if it is readily available on call.

Underestimating the severity of a fire and failing to call for assistance at once is a common fault. It is doubly serious because the result is usually a large fire with heavy losses, instead of a quick stop with relatively minor damage. Apparatus and crews serve no purpose sitting in a fire station. It is far better to call for help and then find out it is not needed than to have to wait until help arrives while the fire makes headway. Our officer takes no chances; the location and apparent size of the fire promise trouble, so help is called for at once.

Size-up does not end, however, when the first-arriving company officer reaches a decision and puts a plan of operation into effect. In the course of any fire, many things can happen. The situation may change drastically and reveal new operative facts about the building, its occupants or contents. When these facts are discovered by or reported to the officer in command, they may cause the officer to review the size-up and make an extensive change in tactics and operations.

*Clark, *Fire Fighting/Principles and Practices.*

Returning to our example, when our first-arriving officer completed the size-up, tactics had to be decided upon to use to fight the fire. As we said earlier, "tactics are those actions that will keep loss of life and property to a minimum." Let us now consider what these actions are. As the late Lloyd Layman said:

> A major fire presents a complex operational problem and this is especially true of all major fires involving buildings or other structures. . . . Tactical procedure on the fire ground can be divided into fairly definite parts and this appears to offer the only logical and practical solution to the problem of studying and teaching fire fighting tactics.*

Layman listed what he called the "Basic Division of Fire Fighting Tactics," in the following order of importance:

Size-up

1. Rescue

2. Exposures

3. Confinement

4. Extinguishment

5. Overhaul

A. Ventilation

B. Salvage

Ventilation and/or salvage operations may be required at any time following initial size-up.

For many years, fire departments in the eastern part of the United States have used 15 basic points to analyze the fire situation. We mention these points here also, even though we discuss the 8 points of "Basic Division of Fire Fighting Tactics" by Lloyd Layman, so that the reader can see the similarity of both methods. The 15 points† are as follows:

1. Location of the fire

2. Extension probability and possibility

3. Life hazard

4. Time

5. Weather

*From *Fundamentals of Fire Fighting Tactics* (Boston, Mass.: National Fire Protection Association).

†Clark, *Fire Fighting/Principles and Practices.*

6. Construction

7. Height

8. Area

9. Occupancy

10. Access

11. Internal protection

12. Water supply

13. Apparatus

14. Manpower

15. Terrain

Another outline of tactics followed by some departments are the 14 basic rules in the fire fighting strategy, as presented by Emanuel Fried in his text.* He lists these rules and further identifies that there are three logical sequences in fire fighting. They are: Phase 1—Locate the fire; Phase 2—Confine the fire; and Phase 3—Extinguish the fire.

Rescue

Rescue is a very broad term. Rescue may be light or heavy; it may involve life-saving, first aid, resuscitator or heart-lung resuscitation techniques, emergency childbirth, or retrieving a child who has fallen down a well. Specialized rescue techniques are needed to get people off cliffs, search a body of water, or extricate the crew or passengers after an aircraft crash. Here, we are concerned with rescue only as a part of fire fighting tactics.

Tactically, rescue ranks first in fire fighting operations because life is more important than property. If the first-arriving company officer is forced to choose between rescue and letting a fire gain some headway, the officer will favor rescue. But an aggressive attack on the fire, or prompt ventilation, is often the most effective method of rescue.

Rescue may be defined as "any action taken to remove living beings from a burning building, or to extricate them from other kinds of danger, and get them to safety." As a fire fighting operation, rescue may be simple or it may become very complicated. Any sound solution will be the result of using every resource at command intelligently, and this, in turn, depends on a good size-up.

Sizing up a rescue situation takes into account the kind of building, everything known about it, the time of day or night, the kind of oc-

*Emanuel Fried, *Fireground Tactics* (Lexington, Mass.: H. M. Ginn, 1972).

cupancy, and whether the emergency (fire or other) presents a threat to life. Size-up also answers these questions: Are there people or animals inside a building that is on fire? How many of them are there? Must they be rescued at once to prevent loss of life? Can they be rescued with the crew and equipment on hand? Will the fire prevent their escape via stairways or other fire escapes? Are they exposed to danger from heavy smoke or toxic gases or vapors?

Contrary to the popular idea of a fire fighter on a swaying ladder with a child over one shoulder, or people jumping into life nets, most successful rescue operations are not dramatic, except for the people involved. While it is true that ladders are often used to effect rescue and nets are used infrequently, the best method of rescue is the easiest and safest possible.

Where a building has protected, enclosed stairways, little more may be needed than to lead the occupants to safety by this route. Where stairs are not protected, and the fire is above a group of occupants, laying a line to protect them from heat and smoke will make it possible to use the stairs. When stairways are exposed to heat and flame from below, however, another route must be chosen. Elevators, particularly the self-service type, are extremely dangerous. They can become a trap, stuck between floors, requiring special tools and extreme effort to free the occupants.

Fire escapes offer another way to effect rescue. By placing ladders at the ends of the lower platforms and directing streams against fire above or at the sides of the fire escape, people can be taken off rapidly and safely. When conditions permit, another approach is to take people off the lowest level platform and down interior stairways, rather than letting them use the straight drop ladder. The greatest danger of the fire escape route is overcrowding; ice, snow, heavy rain, and high winds also menace those trying to use a fire escape.

When ladders must be used to effect a rescue, choosing the right lengths and seeing that they are correctly placed are vital to success. This applies to both aerial and ground ladders. Scaling ladders are the last rescue measure and, when used to bridge the gap between an aerial ladder and a trapped occupant, these ladders are extremely dangerous. While saving life justifies taking great risks, the risk should be weighed and the attempt refused when there is little chance of success. Nothing is gained if both rescuer and victim are killed or injured by a fall. Experienced officers can testify to many occasions when people in fire buildings managed to get out, without any help, under truly amazing circumstances.

At lower levels, aerial ladders offer an excellent method of rescue; the elevating platform also works well in similar situations but is much slower. (See Chapter 5 for comparison of these two pieces of apparatus.) Portable life nets, held by eight to ten persons, may serve as a last resort for rescue but should not be used where the trapped occupant is anywhere above the third story. Even at lower levels, such as the second story, rescue by net is risky. Both fire fighters and the person jumping can be seriously injured or even killed.

Another neglected aspect of rescue operations is searching for persons in smoke-filled parts of a building. Such a search should be conducted by fire fighters wearing masks equipped with a two-way radio. Search is also called for when fire occurs in such places as a school dormitory. Children are particularly prone to panic and take refuge under beds, in bathrooms, or in closets. They must be found and usually carried to safety. Any building search must be well coordinated so that portions of a building, once searched, are not unnecessarily searched again by other fire fighters. This aspect of search has been consistently brought up during studies of fires with large loss of life as one aspect of failing to use all available manpower efficiently.

Police assistance is a must when large numbers of people are involved in a fire. The area must be kept clear, a path maintained for ambulances and medical aid, and people must be kept from reentering the fire building in an attempt to rescue someone or save property.

Exposures

For the beginning student of fire science, the word "exposures" may have little meaning. It is a fire service term, a sort of shorthand way of saying "protection of exposures." But to understand the meaning of these latter three words, we must first define the word *exposures*. There are several types of exposures but what we are speaking of here is external exposures—those buildings, structures, or parts of the surrounding area to which an existing fire could spread. The number and extent of external exposures depends upon many things, mainly the nature of the existing fire and its size. Common sense tells us that an oil tank fire, in a crowded waterfront district, is a greater threat to its surroundings than a shipping room fire in a fire-resistant industrial plant, located on the edge of town. Similarly, the range of exposures widens when there is a possibility of explosion, on days with high winds, and during prolonged dry spells.

Protection of external exposures ranks second on Layman's list of fire fighting tactics because, if done properly, it represents one step toward the second objective of fire fighting strategy: confine the fire.

In addition to external exposures, there are also life exposures, such as arise when a fire building threatens those around it with dangerous vapors, toxic gases, or explosion. Again, the strength and direction of the wind and other weather conditions considered during size-up become important in deciding what action to take.

Finally, some authorities refer to internal exposures, meaning any part of the fire building other than the area where the fire is located. We will consider this very important type of exposure when we discuss confinement.

Protecting external exposures adequately means surrounding the fire and covering every possible extension route. We may not, however, have to cover all sides of the burning building. There may be obstacles to the advance of flame such as thick, strong masonry walls with no windows; there may be wide streets or open spaces. Preplanning and size-up will usually determine the points where protection is needed, with the most serious exposures getting first attention.

While many buildings have some sort of built-in protection against the spread of fire from adjacent buildings, far too many of these systems are incapable of withstanding flames and radiated heat. Often systems fail due to lack of proper maintenance.

Wired glass is commonly found protecting windows in older buildings. When set in metal frames, it will protect the opening it covers against most fires; wooden frames reduce the level of protection to about 177°C. (350°F.). Metal-covered or solid metal shutters are another method of window protection. Those that operate automatically when a fusible link melts are best; those that must be closed by hand are all too often left open, destroying their value completely. If left open, flames and heat from the fire building may prevent their being closed, leaving an avenue that must be protected.

Outside sprinkler systems, "water curtain" or "deluge" (see Chapter 9), are one of the best methods of protecting external exposures, if in good operating condition. Such systems should be supported through the outside siamese connection. Preplanning will have discovered whether such protection is installed and the extent of its coverage.

Failing protection systems or devices, fire fighters use hose streams to protect external exposures. While the size of the stream depends on the height of the buildings involved and the area needing protection, heavy streams are considered much better for this purpose and, when exposure danger is at its greatest, as in the presence of a very hot fire, heavy streams are a must. An incidental benefit of heavy streams to protect exposures is that the backspray cools the space between buildings enough to allow an attack on the fire building from this direction. In planning how to protect external exposures, the officer in charge of operations should always survey conditions on the downwind side of the fire building unless the fire has already gone through the roof. In that event, the buildings on either side that stick up above the fire building need first attention.

In residential areas, the first concern in protecting external exposures is the kind of roofing on the fire building and nearby structures. Wooden shingles present an extreme danger; tar paper nearly as great. The hazards posed by other roofing materials vary from moderate to none.

Confinement

The next step in fighting a fire is to confine it, if possible, and so prevent its spread to other parts of the fire building. Confining the fire is also the second principle of fire fighting strategy, as we saw earlier. The fact that it is a common objective of both strategy and tactics emphasizes the importance of confinement. This emphasis is deserved, for experience shows that when confinement is possible, the result is a quick stop; when it is not, the result is a major or serious fire with heavy loss.

A modern industrial plant, with fire-resistive construction throughout, offers almost ideal conditions for confining a fire. In fact, if the fire department does not arrive quickly, the plant's automatic sprinkler system may well take care of both confinement and extinguishment.

At the other end of the scale are buildings erected before most cities had building or fire prevention codes. Almost every construction feature provides an avenue for fire spread. Buildings that have been extensively remodeled are also likely to be firetraps, full of nasty surprises for fire fighters. There may be false ceilings and unused ductwork, concealed gas lines, ancient electrical equipment, and passages or channels that will allow the fire to get around attempts to confine it with no warning of its escape.

While preplanning based on inspections should give the first-arriving company officer a good picture of a building's structural features and how he will probably have to use crews and equipment to confine a fire, their actual deployment will largely depend on the location of the fire. This is because of the way fire behaves. First, it will extend itself upward (vertically) wherever it can; next, it will spread laterally (horizontally); and finally, if blocked in the first two directions, it will attack any combustible material below its point of origin. This threat to those above is the reason a fire located in the lower stories or basement of a building is regarded as much more dangerous than one involving only the upper stories or the roof.

As we can readily see, a building with unprotected vertical openings, such as stairways, elevator shafts, dumbwaiters, loading chutes, and light and air wells, offers many possible avenues for fire to spread upward. Indeed, a large percentage of loss of life in fires is the result of extremely rapid travel of fire and smoke through such openings. Confining upward spread in such a building can be a very difficult operation, calling for the use of a number of hose lines, perhaps even heavy stream devices, to block potential avenues of spread. Unfortunately, the first-assigned companies may not have enough personnel and resources to cover every possible route. When this happens, we can be certain that the fire will find any way that remains open and follow it.

When a commanding officer has resources enough, the officer will usually start confinement operations by bringing hand lines in on the floor above the fire. From this point, streams can be directed downward into any vertical shaftway. This maneuver calls for good judgment, for if the fire has gained real headway, the floor above it is no place to station fire fighters and hose. Authorities differ on what type of stream to use in blocking vertical shaftways. Many favor the use of fog or spray, because of more efficient heat absorption. Others argue that solid streams are better, because then a considerable amount of water will end up as a pool at the bottom of the shaft, extinguishing burning material that drops.

In more modern buildings, ventilation and air-conditioning systems, or other types of ductwork may take the place of open shafts. Unless protected with sprinkler systems and fire cutoffs, particularly between floors, these metal runways may offer a fire an easy pathway, both vertically and horizontally.

Once we succeed in blocking the upward path of a fire, our next step in confinement is to prevent its lateral movement. Where a building has interior fire walls and fire doors, closing these doors is a good start. Doing so will help prevent the fire's spread to rooms, corridors, and areas beyond them, the internal exposures we mentioned earlier. More often, there will be little or no internal protection. This means that doorways, windows, transoms, and grillwork are all potential avenues for fire spread; even the walls must be watched to see that the fire does not extend itself by burning through them. Once again, we use streams of water to block openings and cool interior walls, floors, and ceilings.

The nature of the building's contents, particularly the way materials are stored and handled, bears heavily on the success or failure of any attempt made at lateral confinement. Crowded storage figures prominently in case histories of major fires. Poor housekeeping practices also contribute to lateral spread by giving the fire something to feed on. Places where large amounts of combustible materials are stored or used in manufacturing or processing present problems in lateral confinement. The danger becomes even greater when there are flammable liquids, combustible metals, or other hazardous materials on hand.

One exception to normal fire spread, where the major danger is lateral movement, deserves mention. This is the situation that faces men fighting fire in a pier warehouse or cargo shed. In general, these structures are poorly constructed. Few, if any, have sprinkler systems; most have no interior partitions, let alone interior fire walls. The contents are usually combustible materials, piled nearly to the ceiling, without space to maneuver between them. Access from the land side is usually limited. When a fire starts, convection and the usual waterfront breezes serve to whip the flames and drive them laterally down what becomes a narrow wind tunnel. Confinement becomes extremely difficult, if not impossible. A wharf differs from a pier in that it parallels the shoreline, which provides increased access. Notice though that in case of fire aboard a ship, just the reverse is true; the bulkheads and steel sides of the vessel restrict lateral spread while open cargo hatches and ventilators favor upward movement of flame and smoke. Shipboard fires do move laterally, due to the conductivity of metal bulkheads separating compartments, holds, and so forth. Therefore, lateral spread must be anticipated and cargo adjacent to these bulkheads moved to prevent ignition. There are six areas of every fire that must be considered in confinement: the top, the four sides, and the bottom.

Extinguishment

With *extinguishment* we come to the crux of all fire fighting tactics: what do we do to put the fire out? The instinctive answer, of course, is pour water on it. In most cases, this is the right answer. Unfortunately, it does not go far enough. We cannot just "pour" water on a fire of any size; it takes people, apparatus, and equipment to supply solid or fog streams in the quantities needed. In addition, there are a good many situations in which water is the last thing to use as an extinguishing agent.

The tactics of extinguishment consist partly of how to get enough water, or some other extinguishing agent, on a fire; partly of how much extinguishing agent to use and where to apply it. What we will do depends primarily on what is burning, how much is burning, and the heat of the fire itself. Preplanning may warn a commanding officer to expect plastics, flammable liquids, rubber, wood turnings, flour dust, and so forth, in addition to ordinary solid combustible materials in a given fire building. But, it is the size-up that lets the officer form a judgment as to what is involved in the fire, how hot the fire is, and how far it has progressed. Taken together, these two sets of facts form the basis for the tactical decisions.

When a fire is confined to the interior of a building, as we have just seen, putting out the fire without much more damage will not be too difficult. Nor will it be hard to supply the amount of water needed to extinguish any burning material completely. Hand lines, with fog or solid nozzles, can readily be advanced to deliver water on the seat of the fire. To avoid water damage, trained fire fighters use only as much water as will serve to put out the fire.

Getting water on a fire when the entire building is blazing fiercely calls for an entirely different approach. Heavy streams from monitors, turrets, aerial ladders, and elevating platforms may be the only way to get enough water on the flames. It may be impossible or useless to try and enter the building and attack the fire from this point.

Fire in enclosed spaces, such as basements, presents special problems in extinguishment. In a basement fire, every effort must be made to prevent its spread to upper floors, yet access may be so restricted that hand lines cannot deliver enough water to check the fire, much less extinguish it. The first line should be brought in on the floor above the fire and used to confine the fire to the burning area, thus preventing its vertical extension. If there are no cellar windows or openings through which special or regular nozzles can be extended, we may have to cut holes in floors or walls to get at the fire. Basement fires may also require personnel to use self-contained breathing masks, something of a handicap, and restrict our choice of stream to fog because a solid stream cannot be directed accurately enough against the seat of the fire to be efficient.

Cocklofts, commonly found in older dwellings, row houses, and tenements, extend across the top of several buildings between the roof and the ceiling of the top story. With access limited to ladders and trapdoors from below, or skylights and shaftways on the roof proper, getting hand lines into position can be extremely difficult. If a fire is well advanced in this type of building, there is great danger that the roof will collapse. A prudent commanding officer orders fire fighters off the roof and resorts to ladder pipes or other heavy streams from above. The officer will also be sure that all fire fighters are withdrawn from the building before large amounts of water are applied to floors above where they were working.

Structural features affect extinguishment tactics to a considerable extent. We have already mentioned the cockloft, found in many buildings in older sections of cities. Table 6 lists some other major construction features of various types of buildings.

Looking at Table 6, we can readily see that a different approach must be used to get water on a fire in a frame tenement than when fighting a fire in a fire-resistive building.

Frame tenements still exist, fortunately in ever-decreasing numbers. Those that are still with us are found in the poorer sections of large cities and some mill towns. They are true firetraps, built of wood throughout, with open stairs and hallways, light and venting shafts, wooden party-walls, a common cockloft, and wooden beams over the cellar touching end-to-end in the cellar walls. Every one of these features favors rapid extension of a fire. Confining the fire to a single structure is vital. At the same time, extinguishing an existing fire must be pushed as rapidly as possible.

TABLE 6. BUILDING CONSTRUCTION FEATURES

TYPE BLDG.	HEIGHT	DEPTH	FRONTAGE	OCCUPANCY	FIRE ESCAPES	SHAFTWAYS	CELLAR ENTRANCE	STAIRWAYS
Tenement Brick Type 1	3–6 story	15–24 m (50–80 ft.)	6.1–7.6 m (20–25 ft.)	1–4 families per floor	front & rear	sometimes vent shaft	under stairs	wood, open
Brick Type 2	5–6 story	15–24 m (50–80 ft.)	6.1–7.6 m (20–25 ft.)	2–4 families per floor	front & rear	light, air, vent, dumb-waiter	under stairs	wood, open
Brick Type 3	5–6 story	15–24 m (50–80 ft.)	6.1–7.6 m (20–25 ft.)	2–3 families per floor	front & rear	dumb-waiter, interior courts[a]	under stairs	wood, open
Brick Type 4	5–6 story	15–24 m (50–80 ft.)	10–15 m (33–50 ft.)	5–7 families per floor	sides & rear	vent, protected	exterior only[b]	fire-proof, enclosed
"Apartment house"	6–7 story	up to 30.5 m (100 ft.)	up to 27 m (90 ft.)	8–12 families per floor	sides & rear	elevator, light & air	under stairs	open combustible
Frame	2–3 story	15–24 m (50–80 ft.)	6.1–7.6 m (20–25 ft.)	1–2 families per floor	none	light & air, courts	under stairs	wood, open
Apartment Newer	5–8 story	up to 30.5 m (100 ft.)	up to 30.5 m (100 ft.)	6–12 families per floor	front & rear	variable	exterior only	fireproof, enclosed

TYPE BLDG.	HEIGHT	DEPTH	FRONTAGE	OCCUPANCY	FIRE ESCAPES	SHAFTWAYS	CELLAR ENTRANCE	STAIRWAYS
Office Non-fire-resistive	5-7 story	up to 46 m (150 ft.)	11-15 m (35-50 ft.)	10-11 offices per floor	rear & side	elevator, central light & air	under stairs	wood, open
Fire-resistive	less than 45.7 m (150 ft.)	varies	varies	1-4 offices per floor	rear	elevator, light & ventilation	exterior only	resistive but open
Fire-resistive	over 45.7 m^c (150 ft.)	varies	varies	1-4 offices per floor	varies	all enclosed fire-resistive	enclosed fire-resistive	enclosed fire-resistive
Loft Non-fire-resistive	4-6 story	varies	varies	1-2 mfg. or processing concerns per floor	varies	elevator, light, stairs, pipe recess	under stairs	open, wood or iron
Fire-resistive	4-7 story	varies	varies	1-2 concerns per floor	varies	enclosed, protected	enclosed or exterior; fire-resistive	enclosed, fire-resistive

[a] Courts may also be in exterior front or rear
[b] With cellar ceiling of fireproof construction
[c] Over 45.7 m (150 feet), a building must have standpipes, fire pumps that will lift water to the highest story, and usually have no burnable material in its construction

Today there are new wood frame structures being built—particularly condominiums and townhouse apartments—that have the same features that lead to severe fire fighting problems that are found in the older tenements.

Recommended tactics in such structures call for stretching the first line to the head of the inside stairway leading to the cellar. If, as is common, the fire started in the cellar and has not gained too much headway, the crew manning this line can direct a solid or fog stream on the flames until the fire is extinguished. If the fire has gone too far to allow entry into the cellar, the first line is used to protect the stairway and hall and a second line is taken up the stairs to where fire in the other parts of the building can be attacked and extinguished. Fire fighters operating this second line should give special attention to extinguishing fire in all shafts, opening them up wherever necessary.

Fire on the top floor calls for immediate attack, including the pulling of the ceilings to prevent its spread to the cockloft and, via this route, to other buildings in the row. Ventilating the building, by removing roof scuttles and skylights, will help remove heated gases and smoke from the building (see Ventilation, page 224). This will keep heat from building up inside the structure and make it easier to advance hose lines.

These tactics of extinguishment also apply to fire fighting in brick tenements (all four types), and the so-called "apartment house" that is really a tenement. The major difference in the attack pattern used against fire in each of these buildings is how the different kinds of vertical shafts are covered and the somewhat greater latitude for operation afforded by windows that open on inner courts. At the same time, these windows may have to be protected as they offer avenues for fire spread.

Brick tenement, Type 4, because of its fireproof cellar ceiling and no inside stairway to the building, favors confinement of a fire to the cellar where it may be extinguished by manned hose lines or cellar pipes. Extinguishment tactics against fire on the upper floors of a Type 4 brick tenement remain the same as those employed against similar fires in the other types of tenement with one exception: little or no attention need be given to the single protected vertical shaft.

In all the other types of tenement, one tactical maneuver will pay big dividends: if the fire has involved a stairwell, a stream should be put on it at once. The stairway can be saved and extinguishment considerably advanced by this measure. Another tactical maneuver, suited for use against fire in all tenements, is to open up the roof, where needed, when a fire has gained headway in the hanging ceiling below the roof. Also, in all these structures, partitions without fire stops require careful attention; they may be acting as vertical shaftways, carrying fire up into the top of the building. Tactics call for pulling an involved partition from its studs, starting at the top, and introducing a stream or fog nozzle at this point to put out the flame.

Fire fighters can easily put out most fires in tenements or apartment houses with the amount of water a 1½-inch line will carry. Fog streams and a nozzle no larger than one inch seem to work best. A larger line, 2½-inch minimum, should be stretched and kept charged as a safety measure.

Fire in a non-fire-resistive office building calls for the same extinguishment tactics as those used against tenement fires except on a much larger scale. These buildings are quick burners and literally riddled with many vertical avenues for fire travel. The major concern at fires of this kind is rescuing people who are frequently trapped on upper floors when elevators fail and smoke, hot gases, and flame block the stairways. At the same time, speed in getting water on the fire is vital. Everything will burn in buildings of this type, so a quick extinguishment is the only way to prevent a major fire and heavy loss. Once again, remember that it is dangerous and unwise to try to use the elevator(s) in a fire building, but often necessary in order to get crews and equipment to the floor beneath the fire.

Buildings over 46 m (150 feet) high, because of the many restrictions on their construction, were once thought to be relatively safe. The major problem appeared to be reaching the fire. But more experience has indicated that this is not true. The contents burn with such rapidity and emit such huge capacities of heat and toxic flammable gases that whole new concepts in what constitutes "high rise" construction, fire protection requirements, and fire fighting techniques have been established. Indeed, buildings in California in excess of 23 m (75 feet) above grade are now classed as high rise. Some jurisdictions simply state that any portion of a building that is beyond the capacity of the local fire department to reach with ladders or aerial equipment qualifies as a high rise structure, and as such, requires a variety of special fire protection requirements that pertain only to high rise structures.

The evacuation of hundreds of tenants within these large buildings is for the most part impractical. The products of combustion render traditional evacuation routes too risky and time consuming. Therefore, areas of protection are being provided wherein occupants can move either up or down from a threatened floor in a building into a relatively safe refuge inside the building itself within a short period of time. "Compartmentation" is the term used in building codes for this concept.

Elevators have been a continual problem, as their very nature permits the spread of smoke, gases, and heat to upper floors. Heat warps doors, and in general causes many varied types of failure in elevators in such buildings. Therefore, codes do not accept elevators as exits. Fire fighters and tenants alike have been trapped in elevators and, where they have been isolated on the fire floor or above it, fatalities have occurred.

Fires in high rise buildings have been rare, but they have also been significant in terms of the loss to lives and property when they do occur. Fire fighters are often required to carry their equipment up stairs and to work in very hot enclosed spaces, which tend to exhaust them quickly and require additional personnel.

True, buildings over 46 meters (150 feet) in height do require special building construction features, and by their nature do tend to isolate fires into specific areas. These same features make the fire more severe and difficult to fight. Built-in fire protection, such as early detection, evacuation, and suppression equipment, will often offset the problem (taller buildings) that modern technology has given us, and will in time reduce the fire situation to a more manageable size.

A structure that has probably caused more trouble for fire departments than any other kind is the *loft building*. Standing from four to seven stories high, each floor of these buildings is usually occupied by a single concern, generally engaged in manufacturing or processing. Entire buildings of this type, such as those in and around the garment district in New York City, are devoted to work connected with a single industry or trade. Combustible materials abound; there may be even more hazardous processes, such as making plastic products. In addition, these manufacturers are frequently in financial straits, greatly increasing the chances of fire set deliberately to collect insurance money.

Extinguishing a fire in an old loft building is no easy task. The entire fire area will probably have to be covered with heavy streams to prevent fire spread, let alone put the fire out. If this confining tactic is successful, hand lines can be advanced to attack the seat of the fire. Where more than two floors are involved, companies should stay out of the building and no one should be permitted on the roof until inspection, backed by experience, shows that it is safe to work there.

The greater danger in fighting a fire in an old loft building is that rapid burning, aided by open vertical shaftways, will weaken the structure very rapidly and cause it to collapse, particularly when heavy machinery is in the building. For this reason, applying water to this type of fire may have to be restricted to heavy streams or streams directed into the building through side windows. Floor space in loft buildings tends to be long and narrow, making streams from front or rear windows ineffective. Remember, to be effective, 90 percent of a solid stream must be delivered within an imaginary 38-centimeter (15-inch) circle drawn around the seat of the fire.

Another danger particular to loft building fires is the concentration of smoke, heat, and fire on the fire floor. Opening the door to attack the fire allows this pent-up fury to escape in the direction of those operating a hand line. Ventilation, through scuttles and skylights on the roof and the overhead door to the stairwell, will greatly reduce this threat to the safety of the fire fighters or even eliminate it. Similarly, because of the danger of backdraft, when a loft building cannot be ventilated quickly through openings on the roof, the situation calls for breaking the windows with a heavy stream.

Loft buildings frequently contain large amounts of baled combustible materials. These materials will absorb large quantities of water, even if wetting agents are used to extinguish the fire. In fact, they will absorb so much that if more than one floor of the building is involved and the fire is not put out in less than an hour, the weight of the absorbed water may cause these floors or the entire structure to collapse.

Remember, however, that floor collapse should be anticipated whenever large quantities of water are applied and little run-off is observed, and in particular where heavy machinery is located, or where fire has damaged structural components. Floors have collapsed in much less than one hour in such circumstances.

In newer loft buildings, noncombustible or slow-burning structural materials reduce these hazards materially. Collapse becomes a minor, not a major, hazard and sprinkler systems will tend to confine fire to where extinguishment is relatively easy.

Large fires in any structure call for the use of heavy streams to put out the fire. Small streams cannot do the job because they do not have the range nor the cooling effect needed. A big fire creates a tremendous draft that breaks up a small stream and keeps most of the water from reaching the seat of the fire.

Direction of attack on a fire is important because when those manning a line approach an open fire from downwind, they may be able to stop the fire at this point. If there is enough heat, acrid smoke, or toxic or nauseating fumes, they may have to attack from the windward side, a disadvantage because the fire is, in a sense, going away from them. A good compromise, if possible, is to attack from the flanks, particularly with heavy streams. Inside a building, unless conditions are against it, the attack should be made from the windward side only because the fire will be stopped by the exterior walls. Such an approach spares those manning a hand line a lot of punishment.

Fire in ordinary combustibles occurs in a number of other types of buildings. The tactics of extinguishing fire in warehouses, theaters, places of public assembly, churches, windowless buildings, lumberyards, on piers, aboard ship, in grass, brush, and forests are described in detail in books on fire fighting tactics and strategy. For this reason, we will limit our discussion of how to put out these fires to one comment: The basic problem remains that of getting enough extinguishing agent on the seat of the fire to put it out quickly. In other terms, we can say that sufficient extinguishing agent applied to absorb all the heat generated by combustion lowers the temperature of the burning materials below their ignition temperatures, thereby extinguishing the fire.

Bulk Flammable Liquid Fires

So far, our discussion of the tactics of extinguishment has concentrated on fires that can be put out with water. But, as we saw in Chapter 6, some fires require special extinguishing agents. One fire of this type, that involving bulk storage of flammable liquids, deserves special comments. Although bulk storage fires are rare, when they do occur they present us with a set of special problems including the threat of a true conflagration.

Usually, tank farm (bulk storage location) fires begin with a "spill" that ignites. A spill, in the form of a stream or pool of flammable liquid, may have its origin in a leaky valve or pipeline, a broken hose, pumping equipment damaged during transfer to tank trucks, a leaky storage tank seam or faulty plate (rare), or even failure of a shutoff device that allows a tank to overflow. Heat from the burning spill steps up production of flammable vapors by the stored liquid, and once the air-vapor mixture reaches the right proportions, trouble really starts.

Fire in bulk storage of flammable liquids means we are facing truly formidable amounts of combustible liquids. A refinery or tank farm may have dozens of tanks, each holding as much as 200,000 barrels of liquid (between 23 and 32 *million* liters, 6.0 and 8.5 *million* gallons); hundreds of 208-liter (55.0-gallon) drums of flammable liquid can be stacked in relatively small areas, such as warehouses or yards. Once a fire starts in bulk storage, it will not lack for fuel.

Size-up of a fire at a tank farm is often very difficult because radiated heat and heavy smoke make it impossible to get close to the fire or to see well. It may even be hard to locate exactly which tank (or tanks) is on fire. At the outset, rescue is a minor consideration in fire fighting tactics at a tank farm fire. It may become important if the fire involves refinery buildings or offices, but even then the number of people working in these highly automated installations is likely to be small.

Protection of exterior exposures probably rates highest tactically in fighting bulk storage of flammable liquid fires because protection forms an essential part of confinement. Protection of exposures itself is so important that if there is not enough water available to cover exposed tanks with cooling sprays, fire fighters must immediately withdraw to a safe distance from the fire, at least 304.8 meters (1,000 feet), and "let'er burn."

Confinement becomes a tricky proposition as well. Preventing spread is a primary objective in fighting any flammable liquid fire, but great care must be taken not to defeat our primary purposes. If we apply large amounts of water to protect exposed tanks, we may cause the burning spill to overflow the dikes that are normally present on a tank farm. This will allow burning fluids to escape from confinement. Yet, if we do not use enough water to keep exposed tanks from buckling or leaking under heat stress, more fuel will be added to the flames and confinement will certainly fail. A reasonable rule of thumb in deciding which tanks to protect and how much water to use in cooling exposed tanks is to use water only on those tanks that create steam when water is applied, and apply it over the entire surface until steaming stops. Repeat as necessary.

Extinguishment becomes the toughest problem of all because there are so many different kinds of flammable liquid and our choice of extinguishing agents must be correct. In addition, because of the extreme heat and the danger of explosion, it will certainly be very difficult, if possible at all, to get close enough to apply enough of our chosen extinguishing agent to put out the fire. Personnel operating lines may have to be protected by a water spray curtain as they approach a burning tank. If so, great care is needed to keep water out of the tank since this can cause the contents to slop over the sides or precipitate a steam explosion that will hurl flaming liquid considerable distances.

The best way to put out an open tank fire, regardless of its contents, is by applying a blanket of the proper foam. As we saw in Chapter 6, certain types of foam are useless against fires in alcohols and other types of water soluble, water miscible flammable liquids. Normally, bulk storage fires involve petroleum products and either mechanical or chemical foam will do the job. Where tanks have built-in foam generators, and the piping is not damaged or destroyed, extinguishment depends on putting them to work. If, as often happens, this built-in protection is lost, those fighting the fire must get close enough to apply a blanket of foam to the surface of the burning liquid.

Another alternative, preferred by experienced chiefs, particularly in fighting oil storage fires, is to drain the contents of the tank until only a small amount remains and then let this remaining fuel burn itself out. If pipelines and valves are undamaged and nearby tanks and installations can be kept cool, this approach is probably the safest. Never forget that even with modern tanks having vents or explosion hatches or floating roofs, a fire in flammable liquids stored in bulk always offers the danger of an explosion.

Needless to say, fires where gasoline, kerosene, or other low flashpoint petroleum products are stored represent an even greater threat to the life and safety of fire fighters and other people than do bulk oil fires. Here the main threat is an explosion of the highly flammable vapors these liquids give off at very low temperatures. Cooling the tanks does little good other than to slow down the rate at which these vapors are produced. These and other hazards of storage fires are covered in detail in *Flammable Hazardous Materials,* by James H. Meidl (Glencoe Publishing Co., 1978).

Liquefied petroleum gases, propane and butane, and liquefied natural gas are frequently stored in bulk. Although a very small LP gas fire can be put out with dry chemical, there is no known method or material that will extinguish a large one. Water, particularly heavy streams, should be applied to all containers, piping, vessels, exposed tanks, and any combustible materials in the area. Water should be applied to vapor spaces of exposed tanks to prevent "blistering" and metal fatigue with resulting rupture of the pressure vessel at this point. This will at least help to confine the fire. Frequently, lowering the temperature will allow pressure relief valves to close, shutting off the supply of gas from individual tanks and extinguishing the fire at that point. Another measure that may help, before an LP tank ruptures, is to shut off any supply line to the burning vessel.

Ventilation

This aspect of fire fighting tactics is perhaps the least understood. Ventilation consists of those operations that will bring in fresh air from outside to replace the heated or contaminated atmosphere inside a fire building. As such, it forms an essential part of any tactical operation against fire.

As we have seen, combustion generates hot gases as well as smoke. Following well-known principles of physics, these hot gases will rise and, if confined, will tend to "mushroom," or rise to the highest point of the building, exerting their heating effect on everything they touch. Figure 7-4 shows how hot gases expand and eventually fill an entire room. The same process can take place inside an entire building. When temperatures go high enough, more of the structure starts to burn. As a result, unless these hot gases are removed quickly, both rescue and confinement operations may be impossible. Similarly, ventilation will help the officer in command to locate a fire accurately and determine its extent.

Ventilation is generally considered to be the work of ladder companies but it may be carried out by other units. Modern practice is to train all personnel in ventilation techniques. These techniques consist of using natural means of ventilation, such as doors, windows, bulkheads at roof level, deadlights, openings in floors or roofs (sometimes supplemented or enlarged by manmade openings), and mechanical means, including portable fans, smoke ejectors, exhaust systems, and fog streams. If no danger to life exists, ventilation should be coordinated with the work

FIGURE 7-3. A modern rescue tool, Jaws of Life, in operation.
Courtesy of the San Francisco Fire Department. Photo by Gregory
Owyang

of those operating the hose lines because premature venting can cause
a fire to spread rapidly. While a potent weapon for reducing the danger
to life, ventilation cannot always be employed effectively and in time
unless the fire is favorably located or an opening in the roof can be
made quickly to draw heat and smoke away from endangered persons.

In buildings without windows or with iron shutters over the windows,
any considerable fire will produce conditions conducive to a smoke explosion or a backdraft. The lack of ventilation allows heated gases and

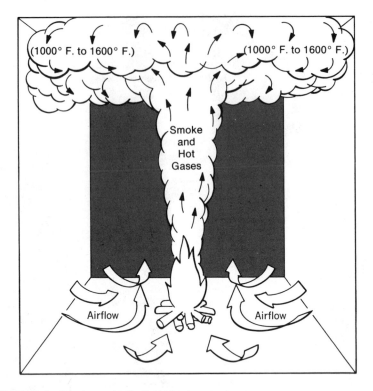

FIGURE 7-4. "Mushrooming" effect of hot gases.

suspended particles of fuel to build up to the point where there is not enough oxygen to support combustion. Then, as soon as a door is opened, as we saw in our discussion of fires in an old loft building, ignition takes place instantaneously. Doors and windows have been blown out, roofs lifted off by the resulting blast. Needless to say, many fire fighters have been seriously injured or killed. Ventilation, usually through roof openings, can prevent this.

Fire escapes often offer a quick way to open up a building and ventilate it. Working from the top down, fire fighters open those windows that face onto the fire escapes. Care must be taken, however, to avoid being trapped by a fire or hot gases surging upward from the lower floors.

Overhaul

Although in last place on Layman's listing, overhaul is in that position only because of a logical process. Major overhaul cannot begin until a fire is out. But overhaul is a most important part of fire fighting.

FIGURE 7-5. Fire fighters ventilating a building filled with smoke and hot gases. Courtesy of the San Francisco Fire Department. Photo by Gregory Owyang

It is the only way to be sure that a fire is completely extinguished and to put the building in as safe and habitable a condition as possible before returning it to its owner's control. During overhaul, fire departments get their best chance to determine the cause of the fire and where it began; it also reveals evidences of arson.

Overhaul generally takes more time than actual extinguishment. It is tedious, dirty, and dangerous work. Many fire fighters have been killed or injured during overhaul because emotional and physical fatigue following the actual fire fight contribute to accidents and foster carelessness. Although there may be no visible flames, personnel may still be exposed to toxic smoke or fumes, dangerous stored materials such as chemicals, acids, and explosives, and the dangers inherent in a generally weakened structure. Persons engaged in overhaul should wear extra protective clothing, including metal insoles and toe caps in their boots, goggles or face shields, and heavy fire-resistant gloves. Properly conducted overhaul is a true team operation because the work must be shared by the entire department.

The actual process of overhaul should be systematic, careful, and closely controlled by company officers. In general, it begins at the top floor and works down. As materials are overhauled, they should be stacked near walls or over supporting members as a precaution against floor collapse. Again, because of the danger of overloading the structure, it is better to soak smoldering materials in a container filled with water than to hose them down. Metal garbage pails, bathtubs, and sinks are excellent for this purpose.

Throwing material out the window is a very poor practice. A rain of debris is not only dangerous, but what lands outside must be handled once again to clear it away. The best way to remove the debris left after any fire is to put it in large garbage cans. Once filled, these can easily be carried outside and placed where they will not interfere with further operations, movement of apparatus, or the public.

Dumping the damaged contents of a building on the sidewalk is also a bad practice. Doing so suggests a lack of concern over the loss suffered by the owners of these goods. The impression it gives the public is that fire fighters are reckless axe-and-water-users whose chief joy in life is destroying other people's property. Careful disposition of damaged goods will do much to dispel this idea.

Salvage

Salvage includes those operations that protect buildings and their contents from unnecessary damage by smoke, water, or other substances used in fighting fires. Properly conducted, salvage improves any fire department's public relations; it also benefits the public by reducing fire losses.

Salvage operations embrace many activities. Goods that might be damaged by fire, water, or smoke are covered or removed. Property saved from the fire is gathered up and protected against further damage. Excess water is diverted and removed from floors, stairways, and other parts of the building. Salvage may also include putting temporary covers on roofs, holes in walls, and over broken windows; cleaning out the inevitable debris of fire fighting and overhaul; covering and protecting exposed furnishings and drying polished furniture that water or chemicals may have reached; and, at times, heating and drying out the building itself.

Preplanning also extends to salvage operations. Inspection will show where fire and water are most likely to spread, the danger to a building's contents from water, fire, or smoke, and the probable effect water will have on a building's structural strength.

Salvage work should begin as quickly as possible, starting on the fire floor if possible. If not, work should start on the floor or floors below the fire unless the fire is on the ground floor or in the basement. While a coordinated attack on a fire, using modern methods such as fog streams and wetting agents, will greatly reduce the amount of water needed to put out the fire, salvage operations below a fire floor can be dangerous. Personnel will be working in a certain amount of heat and smoke and beneath the weight of water-soaked goods, ceilings, and floors.

Salvage operations below the fire floor begin with spreading waterproof covers and fixing them in place. These covers are generally made of heavy waterproofed canvas, although several types of plastic salvage covers are now in use. Standard canvas covers come in two sizes: 3.7 × 5.5 meters (12 × 18 feet) and 4.3 × 5.5 meters (14 × 18 feet). They weigh about 13.6 kg (30.0 pounds) apiece. Like most good tarpaulins, they also come with metal grommets spaced around their edges. Goods, machinery, or equipment, protected by one or more of these covers, are fairly safe from damage by water seeping or running down from above.

Fragile merchandise displayed on tables is difficult, if not impossible, to cover unless there is some way to support the salvage tarps above the tables. Details of the fixtures, such as shelving, counters, and cabinets, have a direct bearing on salvage. Shelving that touches a wall will pick up water as it comes down and carry it to all the goods stored on the shelves. Further, since this type of storage cannot be properly covered, whatever is on the shelves has to be removed. Even a large trained salvage crew would need a great deal of time to accomplish this; this time cannot be spent. Similarly, stock piled on cabinets or cases, with no protection against water, also has to be removed before the showcases can be covered. Dust covers over clothing waiting shipment or in storage off the display floor of stores are usually not waterproof and must be replaced with salvage covers. Items lying about the floors have to be picked up and covered or they can suffer damage not only from water but from the feet and hose lines of the fire fighters. Stock piling is often a major obstacle to salvage operations. Material is often piled too close to sprinkler heads and ceilings to allow quick and complete protection by salvage covers (Figure 7-6).

Dewatering, the second major step in salvage operations, can be a major task or little trouble, depending on the severity of the fire and the extinguishing methods used. The more water, of course, the harder it is to remove. Rapid removal is essential to reduce the dripping that will mean keeping salvage covers in place. As a result, this aspect of salvage operations should be extended to the fire floor as soon as possible.

Where there is little water used, as in residential fires, sponges, squeegees, mops, rags, and water vacuums can easily remove it all. But in

FIGURE 7-6. Covering property with salvage covers. Courtesy of the Oakland Fire Department.

multi-storied buildings after a fire of any severity, removal of excess water calls for a variety of equipment. In some buildings, there will be floor scuppers, originally designed for drainage, or floor drains of some type. If these scuppers or drains are not clogged or plugged, as they too often are, they afford an easy method of dewatering an area. Usually, though, water must first be contained by bagging or damming, usually with sawdust, and then diverted to where the runoff will do the least harm. Particularly on the fire floor, the sprinkler system will have to be shut off and any open heads replaced or plugged.

Where concentrations of water are heaviest, salvage crews may have to cut holes in the flooring and discharge the water by chutes or funnels, which are usually made from salvage covers, out of doors, windows, or down shaftways. Ceiling light fixtures may offer an almost ready-made path to discharge water via the floor below. The nearest window is the best place; elevator shafts should not be used. Often sewer drains on the involved floor can be used to divert water. Something to watch for when discharging water from upper windows is recessed basement windows that will catch the water and carry it into the basement.

While it is best to discharge excess water directly from the fire floor to the outside, the ceiling below may be so weakened that, to prevent collapse, it may be wise to puncture it and divert the water above into prepared channels. Stairway drains can be very useful in dewatering operations as this is a route usually taken by excess water. The important thing here is to close hallway doors and all other openings so that the water runs all the way down the stairs and out the ground floor exit. Catchalls or basins, also improvised from folded storage covers, can impound water dripping from above and also can be used to store wet, salvageable material in a way that protects it from further damage.

A major problem in salvage operations is a flooded basement. Drains may be blocked, sewer and waste pipes inaccessible. If not removed quickly, water in a basement can knock out the building's heating plant, force shutting off its electricity and elevator service, and spread into stockrooms, damaging their contents seriously. It can also seep or leak into the basement of adjacent buildings. For these reasons, salvage crews or companies should have quick access to pumping equipment.

Basement flooding is also inherently dangerous to fire fighters. Natural gas pilot lights may have been extinguished, liberating the gas into the atmosphere, and submerged electrical equipment such as motors, pumps, and outlets, may have energized the water without causing a circuit overload and circuit safety features to operate. Hence, overhaul in these areas can be very dangerous. Adequate consideration should be given to these dangers through preplanning and size-up, particularly when time is available and the emergency is past.

Covering and enclosing openings made in the structure by fire or ventilation are part of salvage operations. Holes in roofs must be taken care of promptly if the protection given to the building's contents is not to be wasted. Building papers and light wood frames are the usual method of covering holes in roofs, replacing window glass and missing doors. Large openings call for the ever-present salvage covers. Care must be taken in working on roofs since supporting timbers may have been badly weakened by fire and collapse under the weight of those trying to repair them.

Covering and enclosure are particularly important in winter. Ice formation, with subsequent melting and dripping, can cause great damage; in places with perishable goods, freezing will result in heavy loss. Once the fire building is enclosed, portable heaters can be brought in and used to prevent structural and content damage by speeding drying out and keeping the building warm.

The hallmark of good salvage operations is that they reduce the amount of time and money lost because of fire.

REVIEW QUESTIONS

1. What is meant by tactics in fire fighting?

2. What is meant by strategy?

3. What constitutes size-up and how is information obtained to accomplish it?

4. Name the five parts of a size-up.

5. How is preplanning used in determining strategy and tactics?

6. The knowledge of where a burning building is located is important when determining the fire ground strategy because _____.

7. Name 8 of the 12 construction features of a burning building used in determining tactics.

8. What must be known about the occupancy of buildings in determining tactics and strategy?

9. Why should the original size-up and resulting decisions be flexible?

10. Name the basic 15 points used to analyze a fire situation.

11. Define rescue operations. Why is it the first tactic?

12. Name several ways to evacuate people from a building and name the advantages and disadvantages of each.

13. What is meant by the term *exposures?*

14. Name some of the methods used to protect exposures.

15. What is fire confinement and why is it an important objective of both strategy and tactics?

16. Name some of the obstacles to confinement.

17. How is preplanning used in the tactic of extinguishment?

18. Name several structural features of a building that deter extinguishment.

19. Name several structural features that help extinguishment.

20. Name some of the tactical problems in high rise buildings; in loft buildings.

21. Outline the tactical problems of size-up, exposure, confinement and extinguishment in fighting flammable liquid fires.

22. Explain the need for and operation of ventilation.

23. What is overhaul and why is it an important fire fighting tactic?

24. What is salvage, as used in fire fighting tactics? What operations are included as part of this tactical division?

8 Private Fire Protection

So far, we have been concerned primarily with public protection: fire protection furnished by a public agency, such as a city or county fire department. Let us now take a look at another type of fire protection that calls for a knowledge of fire science: planning and setting up private protection (fire protection provided by the owner of property).

This is a field that offers many opportunities to the student of fire science. There are literally thousands of people, trained in fire science, who play important parts in industrial fire safety programs; and as many more concerned with the manufacture, installation, and sale of equipment for use in private fire protection. Since the requirements for adequate fire protection vary so widely with type of occupancy, we will limit our discussion to broad principles rather than attempt to cover every detail.

Organization

How big and what kind of private fire protection organization is needed depends upon the nature and hazards of the property to be protected. A large industrial plant, for example, requires a different type of organization than a school or hospital. But remember: *no property,* including a private home, *is too small* to justify an organized fire defense program.

Management

Since the general manager or top executive officer of a corporation or institution is usually expected to take all proper measures to protect property and life, fire protection becomes a direct concern of top management. Large corporations, with a number of plants located across the country, frequently assign a qualified fire protection engineer to the staff of the president or top officer. Designated as the Director of Fire Safety, Fire Loss Prevention Manager, or some similar title, this person works directly with top executives in setting up the company fire prevention system and program. Part of the job is to develop organizations for fire protection within each plant that fit in smoothly with the work of the different departments and also conform to general company policy. To do this calls for skill, care, and a good measure of tact.

Other responsibilities of this position include making a comprehensive analysis of possible lost earnings because of damage to company property, personal injury or forced layoff, and interruption of production. Such an analysis gives management an idea of potential loss from fire, its probable effect on the company, and why money spent for fire protection is well spent.

Some large corporations have what is called "decentralized organization": each division operates as if it were a completely independent company. In this type of operation, division or plant management is responsible for fire prevention measures.

When an organization has only one plant or facility, the plant manager is responsible for plant protection measures. In large plants, the manager frequently has a staff assistant, trained in fire science, to advise in how to prevent fire.

To be effective, a plant fire prevention manager should rank with the plant engineer or similar division or department heads. Since anything a plant does, has, uses, or makes may affect fire prevention, such a manager's work will cross all organizational lines.

Diplomacy, as well as thorough professional knowledge, is essential. The fire prevention manager must be able to sit down with all department heads and explain the need for fire prevention measures and must be able to work out safeguards that will not interfere with production or normal operations. As a fire science professional, the fire prevention manager is responsible for recommending, designing, and supervising plant fire protection systems and equipment, and for organizing a plant fire brigade and seeing that it is properly trained. The fire prevention manager should help to establish a close relationship with the public fire department and work out mutual assistance agreements with other plants. Establishing guard services and the necessary schedules for inspections are also up to the fire prevention manager. As we can readily see, the post of plant fire prevention manager is an important one.

.Private Fire Brigades

In simplest form, a company fire brigade consists of certain employees assigned to definite fire prevention tasks and responsibilities. In most establishments, a brigade is a very simple organization with the manager (or an assistant) in charge and each employee-member having a specific assignment in case of fire. These assignments may include handling fire extinguishers or small hose lines, shutting down certain equipment, or directing the safe and orderly evacuation of an area or building. In large plants, organization is necessarily more complex; the private fire brigade in very large establishments resembles a municipal fire department.

The fire prevention manager or the plant manager fixes the size and character of the fire brigade, provides supplies and equipment, sees that it is suitably staffed and trained, and selects fire brigade officers.

There should be a chief of the brigade. Part of this job is making periodic inspections of fire fighting equipment, requisitioning supplies, and arranging for equipment repair. The more important part, however, calls for working out plans of action to meet any possible fire situation in the plant, periodic reviewing of brigade personnel, recommending additional members, suggesting changes to the fire prevention manager, and planning and conducting the training program. Enough assistant chiefs should be appointed to guarantee a chief officer on duty at all times.

In very large plants, the brigade chief and equipment operators have no other duties, but usually members of the fire brigade are regular employees. They should be chosen with care from people qualified by normal responsibilities and knowledge of the plant, not merely for availability. Careful scheduling will prevent conflict of duties and cover absences such as off-duty periods, sickness, and vacations. There should be minimum physical requirements for brigade members, especially those who might be involved in fire fighting. Each member of the brigade should have a distinctive form of identification to assure free movement about the plant.

To have an effective fire brigade, all members should have to complete a training course and training sessions should be held at least monthly. The instruction program should cover handling of all fire and rescue equipment in the plant, such as use of portable extinguishers and hose lines, ventilation of buildings, and salvage and rescue operations. The chief should direct practice drills at regular intervals to assure smooth teamwork. These drills should include operation of extinguishers, hose lines and protective equipment, as well as the use of all fire fighting apparatus.

While the size and character of a particular plant will determine the equipment it needs, all plants need portable extinguishers of the proper type. NFPA Pamphlet No. 27, "Private Fire Brigades," has a helpful list.

Employee Participation

To be successful, fire prevention programs must have wide employee participation. For safety, every employee should know what to do in case of fire; for that extra measure of care and awareness of fire hazards, all employees should feel they have a part in the company fire prevention program.

Arousing general interest should not be too difficult because the normal operation of a fire prevention department brings its members into contact with people in every other department. Before equipment goes in or new construction begins, the plant's chief engineer might discuss plans with the fire prevention manager to avoid creating unnecessary hazards. Similarly, there is a link to the maintenance department because facilities and structures must be kept in a condition to minimize the chance of fire. Ideally, the production department should supervise plant operation in such a way as to prevent fire; in practice, this department is often the greatest source of trouble for those concerned with fire prevention. Good top management, however, will insist that all departments participate in fire prevention, seek and follow the advice of the fire prevention department, and observe established company rules concerning operations and possible hazards.

This close association with all departments gives the fire prevention manager an opportunity to spread the word about fire and fire loss. Sometimes this is done through a program of competition between departments much like safety campaigns against personal injury. For smaller plants, a common approach is to assign each employee some responsibility for fire prevention.

Guard Service

The plant security manager or fire prevention manager is usually responsible for guard services because the functions of such services fall under both departments. These services generally consist of (1) controlling the movement of persons within the property, (2) carrying out procedures for the orderly conduct of some operation at the property, and (3) protecting the property when management is absent.

Guards may be plant employees or employees of an outside firm, such as the Burns or Pinkerton agencies. If guards come from outside, the firm's policies for screening, training, and supervision of guards should be closely checked.

Even the best guard system, however, requires supervision, especially supervision of patrol personnel who make regular rounds to see that there are no unauthorized persons on the premises, and that nothing unusual is happening on the property. To be effective, any guard service must have a way to communicate both inside and outside the property. Whether the system uses telephone, telegraph, or radio, design and operation should minimize the chance that service will be interrupted.

There are several types of approved supervisory systems. The most common is the clock and time recording system. Under this system, patrol personnel follow established routes that cover, or are at least in sight of, an entire portion of the plant. Stations along the route are so located and numbered that the patrol personnel must walk or ride to the farthest part of their assigned territory in order to make all stations in the proper order. Each station has a key. When this key is turned in the special clock carried by the patrol person, a disc within the clock records the station number and the time. By looking at these discs, management can tell where the patrol personnel were at all times during their tours of duty. To supplement the recording clock, the stations often have a signal device that will record the time and station on a central recorder at the plant's guard control center or at the guard company's central station.

Guards should be able to call the guard control center and the center should have communication links with the outside. In large plants, the control center should be staffed at all times and should have a directory of names and telephone numbers to call in case of emergency, as well as other pertinent information. Even in small plants, some form of guard control center is important.

There are other forms of guard supervisory service; some include runners who respond whenever there seems to be a break in service. This "back-up" may be provided from the guard control center in large establishments. Smaller plants, in large population centers, usually contract with a company specializing in guard service. These companies maintain a central station, staffed around the clock, that is equipped to receive signals from many sources, including guard tour signals and burglar alarm and automatic fire alarm systems.

For a detailed discussion of guard service, see "Guard Service in Fire Loss Prevention," NFPA Pamphlet No. 601.

First Aid Fire Protection

"First aid" fire protection is the do-it-yourself type. It should only be used *after* calling the fire department, if one is available. But, since most fires are small to begin with, many can be extinguished by someone on the scene who has something to work with and knows how to use it. This is what makes first aid measures an important factor in private fire protection.

Simple Aids

An early first-aid appliance, still in use in some textile mills and yard storage in isolated locations, is the water barrel and buckets. This consists of a 208-liter (550-gallon) barrel, filled with water (protected against freezing in cold countries) and three or more 11-liter (12-quart) buckets.

To operate: fill buckets from the barrel; throw water on the fire. Horizontal reach is limited to about ten feet, but the results can be very effective if the fire is not too large.

The sand bucket, another familiar first aid device, is rapidly being replaced by more sophisticated and effective equipment. Its use is limited almost exclusively to small flammable liquid fires in pans or other small enclosures. Sand should not be used where any moving parts are involved in a fire.

Fire blankets, that can be folded and stored in convenient locations, measure about six feet by eight feet. They are made of woven asbestos, aluminized fabric or flameproofed wool and can be used to smother small fires in flammable liquid tanks, etc. Blankets can also be used for other purposes, such as covering flammable material during welding or cutting operations, as a shield against sudden flame, and to save a person whose clothing is on fire from severe or fatal burns.

The ordinary garden hose is also an effective first aid fire protection device. It is usually readily available in the home and, in some industrial plants, charged garden hoses are kept ready to protect especially hazardous operations. Ordinarily, garden hose nozzles limit discharge, but there are others, designed for fire protection service and approved by Factory Mutual Engineering Division. These special nozzles can be set for various spray patterns and straight streams. By test, the best these nozzles will do with hose pressure at 344.5 kPa (50.0 psi), is a straight stream of 18.5 lpm (4.9 gpm) and a flat spray of 23 lpm (6.1 gpm); at hose pressure of 689 kPa (100 psi), straight stream delivery goes up to 26.5 lpm (7.00 gpm) and spray delivery to 33.6 lpm (8.90 gpm).

Portable Fire Extinguishers

The National Fire Protection Association's "Standard for the Installation of Portable Fire Extinguishers" (NFPA No. 10) states:

Fire extinguishers can represent an important segment of any overall fire protection program. However, their successful functioning depends upon the following conditions having been met:

1. The extinguisher is properly located and in working order.
2. The extinguisher is of proper type for a fire which may occur.
3. The fire is discovered while still small enough for the extinguisher to be effective.
4. The fire is discovered by a person ready, willing and able to use the extinguisher."*

*From "Portable Fire Extinguishers Standard," NPFA No. 10, copyright 1975, National Fire Protection Association, Boston. Reprinted by permission.

Properly located extinguishers will be out in the open, where they can be seen and reached easily by anyone. Where floor area is large, painting supporting columns or walls a distinctive color helps people to spot extinguishers quickly.

Recognized testing laboratories, such as Underwriters' Laboratories and Underwriters' Laboratories of Canada, classify portable extinguishers according to proposed use and relative effectiveness against a standard test fire.

There are four basic types of fires, as follows:

Class A—fires in ordinary combustible material, such as wood, cloth, paper, rubber, and many plastics.

Class B—fires in flammable liquids, gases, and greases.

Class C—fires in electrical equipment with current flowing.

Class D—fires in combustible metals, such as magnesium, titanium, zirconium, sodium, and potassium.

Every approved extinguisher has a label with one or more of these letters, indicating the *type of fire* against which the extinguisher may be used safely and effectively. For example, a C label simply means that the extinguishing agent is a nonconductor and can be safely used on energized ("live") electrical equipment (if the current is off, Class A or B extinguishers can be used); a D label extinguisher would be safe and effective on combustible metals but not necessarily against an electrical fire. Details regarding use are usually given on the nameplate. Read them!

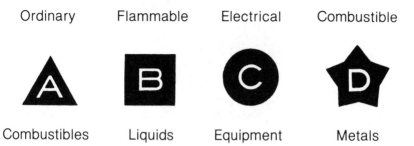

FIGURE 8-1. Markings for extinguishers indicating the types of fires for which they are intended. Background colors are green for Class A, red for Class B, blue for Class C, and yellow for Class D.

When a number *precedes* the class letter, it shows relative effectiveness; the larger the number, the larger the fire (as compared to a standard test blaze) the extinguisher can handle.

For example: Underwriters' Laboratories labels one extinguisher as 4A: 16B: C. This means it will extinguish twice as much Class A fire as a 9.5-liter (2.5-gallon) water extinguisher, rated 2A, that puts out the standard test fire. It also means the extinguisher will put out a flammable liquid fire in a pan 4.9 m (16 feet) square. This is sixteen times as much as a 1B extinguisher, used by an experienced operator, can handle. It also shows that the extinguishing agent is a nonconductor and can be safely used on live electrical equipment (Figure 8-2a and 8-2b).

FIGURE 8-2a. Hand cartridge type extinguisher. Courtesy of the Ansul Chemical Company

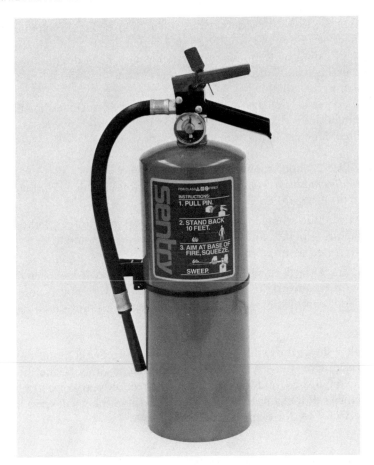

FIGURE 8-2b. Stored pressure type extinguisher. Courtesy of the
Ansul Chemical Company

Selecting the proper extinguisher for private protection depends primarily upon the hazards present. Are they flammable liquids, molten
metals, plastics, chemicals, electrical equipment, or ordinary combustibles? But there are other considerations, including the need for lightweight extinguishers when small, old, or very young people will probably
have to operate them, possible corrosion, possible harmful reaction
between the extinguishing agent and the manufacturing processes or
the equipment, vibrations that may call for special brackets, and many
others. Choice of extinguishers must also take into account the health
and safety of the operators. Extinguishers of the vaporizing liquid type
contain an extinguishing agent that releases toxic vapors; these should
never be used or serviced in a small room, closet or other confined

space. In the 1960s the federal government, many states, and many cities no longer permitted the use of the vaporizing liquid type of extinguishers, and their listing by testing laboratories has now been discontinued. Different extinguishers offer different dangers and these should be considered before installation.

The degree of fire hazard of a given occupancy determines the number of extinguishers that a building, or part of a building, needs. There are three degrees: light, ordinary, or extra:

- Offices, school rooms, churches, and places where there is only *a small amount of combustibles* or flammable liquids are rated as the lightest hazard.

- Mercantile storage, auto showrooms, parking garages, light manufacturing plants, and places where *moderate fires* might be expected are rated as ordinary hazard.

- Extra hazard is where fires of large size may be expected, such as in woodworking and auto repair shops; warehouses and high-piled combustible storage; or flammable liquid handling and storage.

Table 7 shows the extinguisher requirements needed to protect a building, based upon Class A fires. These requirements may be met by using more extinguishers of a lower rating to protect ordinary or extra hazard occupancies. For example, three Class 1A extinguishers will serve instead of one Class 3A for extra hazard occupancy.

TABLE 7. FIRE EXTINGUISHER DISTRIBUTION FOR CLASS A HAZARDS

MIN. BASIC EXT. RATING FOR AREA	MAXIMUM TRAVEL DISTANCE TO EXTINGUISHER		AREA TO BE PROTECTED PER EXTINGUISHER					
			Light hazard occupancy[a]		Ordinary hazard occupancy[a]		Extra hazard occupancy[a]	
	m	ft	m²	sq.ft.	m²	sq.ft.	m²	sq.ft.
1A	23 m	75 ft.	278.7	3000				
2A	23 m	75 ft.	557.4	6000	278.7	3000		
3A	23 m	75 ft.	836.1	9000	418.1	4500	278.7	3000
4A	23 m	75 ft.	1045.1	11250	557.4	6000	371.6	4000
6A	23 m	75 ft.	1045.1	11250	836.1	9000	557.4	6000
10A	23 m	75 ft.	1045.1	11250	1045.1	11250	836.1	9000
20A	23 m	75 ft.	1045.1	11250	1045.1	11250	1045.1	11250
40A	23 m	75 ft.	1045.1	11250	1045.1	11250	1045.1	11250

[a] See Appendix C for degree of occupancy hazard.

Table 8 covers the distribution of extinguishers for Class B fires. Except for foam extinguishers, these requirements cannot be met by using more extinguishers of a lower rating. Larger extinguishers may be used to meet requirements, provided travel distance does not exceed 15 m (50 feet).

TABLE 8. EXTINGUISHER REQUIREMENTS FOR CLASS B HAZARDS
(for extinguishers labeled between 1955 and 1969)

TYPE OF HAZARD	BASIC MINIMUM EXTINGUISHER RATING	MAXIMUM TRAVEL DISTANCE TO EXTINGUISHERS	
Light	4B	15 m	50 ft.
Ordinary	8B	15 m	50 ft.
Extra	12B	15 m	50 ft.
(For extinguishers labeled after June 1, 1969)[a]			
Light	5B	9.1 m	30 ft.
	10B	15.0 m	50 ft.
Ordinary	10B	9.1 m	30 ft.
	20B	15.2 m	50 ft.
Extra	20B	9.1 m	30 ft.
	40B	15.2 m	50 ft.

[a] Note: This complies with the new UL rating system.

   After a portable extinguisher system is properly installed, it must be properly maintained. This means regular inspections, at least monthly, to ensure extinguishers are where they should be, have not been actuated or tampered with, and to detect any obvious physical damage, corrosion, or other impairments. Maintenance includes recharging when required and hydrostatic tests any time an extinguisher shows evidence of corrosion or mechanical injury. Most types of extinguisher should undergo a hydrostatic test at least once every five years, regardless of appearance, although dry powder and gas types may safely go for twelve years.

Standpipe and Hose Systems

Properly installed and maintained, a standpipe system provides inside protection to buildings. It can be used as a first aid device by the occupants and by the fire department for effective inside fire streams, particularly on upper stories.

   The National Fire Protection Association's "Standard for the Installation of Standpipe and Hose Systems" (NFPA No. 14) groups standpipe systems according to three classes of service: Class I service can furnish the fire department with effective fire streams for advanced stages of a fire inside the building or an exposure fire. Class II service offers the occupants of a building a way to control incipient fires. Class III service is a combination of the other two, capable of furnishing effective fire streams or a way to fight incipient fires.

Standpipe systems may have several different kinds of water supply. The most common type is the *wet standpipe system.* Here, the supply valve to the system is open and water pressure is maintained at all times. Three types of systems are used in unheated buildings in areas where freezing weather is common: one is so arranged that when a hose valve is opened, water is admitted to the system automatically through the use of approved devices; the other type admits water to the system by manual remote control devices at each hose outlet. The fourth type is the *dry standpipe.* This system is not connected to any water supply. Its purpose is to give the fire department a quick way to put hose lines into operation on the upper floors of a building. Fire fighters connect their hose lines to the standpipe at the floor involved; water comes from a pumper connected to the standpipe system at ground level.

For Class I and Class III services, standpipe systems should be designed for a minimum flow of 1890 liters (500 gallons) per minute. Where more than one standpipe is required, design flow is determined on the basis of 1890 lpm (500 gpm) for the first and 946 lpm (250 gpm) for each additional standpipe. Piping itself should measure no less than 10 cm (4.0 inches), interior diameter (ID) for a standpipe rising less than 30 m (100 feet), and not less than 15 cm (6.0 inches), ID, if the system rises higher than that. Very high buildings, over 83.8 m (275 feet), must have specially designed systems with separate supplies for upper stories. Systems for Class II are designed to supply a minimum flow of 380 liters (100 gallons) per minute; piping should be no less than 5 cm (2 inches) ID to serve 15.25 m (50-foot) heights and no less than 6.4 cm (2.5 inches) for greater heights.

Supplying water for buildings of fifty to one hundred stories may appear to be a difficult problem. Actually the same principle is used for most buildings over 83.8 m (275 feet) high. The building is divided into pressure zones of about twelve stories for each zone. Fire pumps and gravity storage tanks are installed at each zone level except the lowest level where water may be supplied from the public water system. The pumps take suction from the supply tanks and pump the water to the next higher level until all zones are supplied. There are variations to this method, such as supplying high-pressure pumps to increase the zone limits, but this requires pressure-reducing valves at hose-valve outlets and other complications. The size of the tanks must be calculated according to fire protection demands and domestic consumption if the system is used for both. Sprinkler supplies may also be provided by the system.

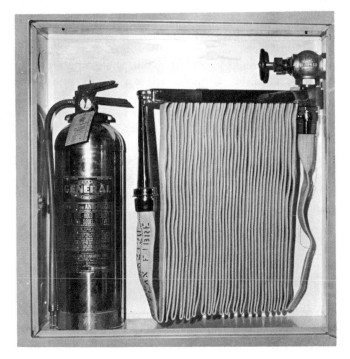

FIGURE 8-3. Hose rack for Class III standpipe system. Courtesy of
the Palo Alto Fire Department. Photo by Jim McGee

The number and location of standpipes may vary according to occupancy, type of construction, exposures, etc., but the standard provides
general rules that will answer most requirements.

For Class I and III services, there should be enough hose stations
so that all portions of each story of a building are within 9.1 m (30
feet) of a nozzle, attached to 30.5 m (100 feet) of hose. For Class II
service, all portions should be within 6.1 m (20 feet) of a nozzle, attached
to not more than 23 m (75 feet) of hose. When a building has exposures
within 18.3 m (60 feet) of it, standpipes for large hose should be located
where this hose can be used for both exposure and interior fires. In
general, large hose outlets should be located in stairway enclosures; small
outlets in corridors or spaces adjacent to the stairway. This will make
it possible to use the small hose in case the stairway is being used to
evacuate the occupants.

All hose outlets should be equipped with a valve between the outlet
and the riser. Small hose outlets should have not more than 23 m (75
feet) of approved small fire hose attached to the outlet and securely

racked. Nozzles for small hose should not exceed one-half inch and shutoff nozzles are usually desirable.

Water for standpipe systems usually comes from the public water system through a gate-and-check valve, but where a public water system is not available or is not suitable, a water supply must be designed to meet the proper standards of flow and pressure. All standpipe systems should be hydrostatically tested when first installed and inspected frequently to be sure they are ready for emergency use.

When high buildings, over 30.5 m (100 feet), are under construction, a standpipe system should be installed and made operational well before the building reaches a height above the reach of the longest fire department ladders. This may be temporary or part of the permanent system for the building (Figure 8-4).

Outside Protection

Most establishments depend upon the public water system to supply water for fire protection, but there are many industrial plants and institutions in places not served by a public water system. Some of these have large yard storage or the buildings cover such large areas that a separate water system is needed to supply hydrants and other fire defense equipment. Such a separate system is referred to as *outside protection*. It includes underground piping systems, control valves, hydrants and hose, or other ways to bring water from its source to a fire.

Water Supplies

Designing an outside protection system calls for an experienced engineer because all the problems of a public water system must be met. The first question is: what demands are to be placed upon the system, or how much water and at what pressure should the system be expected to deliver? The answer depends upon fire flow requirements, based upon area, height construction, and exposure of buildings, as well as the fire demands that may be expected from any yard storage. The answer must also take into account supplying automatic sprinkler systems and what proportion of the plant is so protected.

It is highly desirable to keep the fire protection system and the system supplying water for process and human use completely separated. This permits increasing pressure for fire use, if necessary, provides a definite amount of water for fire use only, makes possible the use of supplies not suitable for domestic use, and permits the fire system to be controlled and supervised more readily.

INSPECTION GUIDE
STANDPIPE SYSTEMS

ADDRESS			DATE	
DBA			OCCUPANCY TYPE	
CONTACT	TITLE		PHONE	
BLDG. TYPE	HEIGHT	NOTICE	REFERRAL	
INSPECTOR	ASSIGNMENT		PHONE	

DRY STANDPIPE SYSTEM

NO.	ITEM	CODE	REQUIREMENTS
1.	Fire Department inlet connection	57.01.35 Plumbing Code*	Inlets must be unobstructed and easily accessible. Inlets are to be complete with gasket and protective cap. Swivels must be operable and free from damage. Check valve must be unobstructed and operating freely. **NOTE** — manipulate check valve (clapper or piston type) and visually check for adequate clearance. Inlets are to be adequately protected from damage and properly supported or secured to the building. Each inlet connection must have a proper identification sign plate.
2.	Outlet Valve	57.01.35 Plumbing Code*	All valves must be easily accessible for fire department use. Threads on valves must be free from damage. Valves must open freely (inspect valve seat visually for proper operation). All outlets are to be maintained capped and with the valve closed.
3.	Piping	57.01.35 Plumbing Code*	All piping must be free from damage or corrosion. Exterior pipe supports must be in good condition.

INTERIOR WET OR COMBINATION STANDPIPE SYSTEM

NO.	ITEM	CODE	REQUIREMENTS
1.	Fire Department inlet connection	57.01.35 Plumbing Code*	Inlets must be unobstructed and easily accessible. Inlets are to be complete with gasket and protective cap. Swivels must be operable and free from damage. Inlets are to be adequately protected from damage and properly supported or secured to the building. Each inlet connection must have a proper identification sign plate.
2.	Outlet Valves	57.01.35 Plumbing Code*	All valves must be easily accessible for fire department use. Threads on valves must be free from damage. Valves must be free from leaks, serious corrosion or damage. **NOTE** — Do not operate valves — if visual inspection warrants a test, require the owner to have one made. Pressure restricting orifice plates, where required, shall be maintained behind the gasket in the female hose fitting.
3.	Piping	57.01.35 Plumbing Code*	All piping must be free from damage or corrosion. Exterior pipe supports must be in good condition.
4.	Hose and Hose Cabinets	57.01.35 Plumbing Code*	All hose cabinets must be accessible and properly labeled, FIRE HOSE. 1½" hose must be free from deterioration, rot or cuts and connected to outlet valve. 2½" hose, where required, must be in a locked cabinet and labeled, FOR FIRE DEPARTMENT USE ONLY. All hose must be complete with nozzle and properly positioned in the cabinet. All nozzles are to be free of foreign materials. All cabinets must be free of other than fire fighting equipment.
5.	Water Supply	57.01.35 Plumbing Code*	Public system pressure and discharge must be adequate. (Request responsible person to open roof outlet for visual check.) Automatic fire pump (where installed) must be operable (request responsible person to open drain valve which will activate fire pump). Pressure tank (where installed) must be maintained full at a minimum of 75 psi. Pressure gravity tank (where installed) must be maintained full and free of excessive sediment. All tanks must be free of leaks and corrosion. Outlet valve must be open.

*For final action see Company Standards.

USE REVERSE SIDE FOR REMARKS, COMMENTS, SKETCH PLAN, ETC.

FIGURE 8-4. Example of inspection guide for standpipe systems.
Courtesy of the Palo Alto Fire Department.

In special cases, combined fire and domestic systems are acceptable when all of the following conditions exist:

- Domestic use is small, relative to fire needs.
- The water is clean and noncorrosive.
- There are ways to control domestic lines.
- The expense of separate systems is not warranted.

In such systems, domestic consumption must be added to fire flow requirements in determining the demands of the system.

Once system demands are fixed, the next problem is an adequate source of water, available and reliable at all times. In developed industrial districts, there is usually a public water system able to supply adequate quantities of water. One or more connections to such a public system affords a good primary supply. Adequacy can readily be determined by flow tests, but no outside protection system should depend upon any pipe smaller than a 15-cm (6.0-inch) loop; and larger pipe sizes are usually needed. Connection to a public supply should be through a valve, preferably of the indicator post type, that shows whether the valve is open or closed. If meters are required, they should not interfere with the flow of large quantities of water.

If public water supplies are unavailable or inadequate, another water source must be found. This may be a lake, pond, river, small stream, or even a well, but using these sources will mean installing supply mains, in most cases, to make the water available for fire protection.

Pumps and Tanks

The National Fire Protection Association's "Standards for Centrifugal Fire Pumps" (NFPA No. 20) goes into detail on the installation, general arrangement, tests, operation, and maintenance of fire pumps to assure adequacy and realiability. This standard exceeds one hundred pages and is constantly being revised to keep it up to date. For our purposes, we need only remember that when a pump is the only means of supplying water to an outside protection system, the system should have a built-in warning service that will indicate whether the pump is operating normally or not; there should also be a similar device for monitoring the power supply. Pumps should be run at full capacity at least once a month to assure proper operation in case of emergency.

Sometimes water must be pumped in two stages: from a well or other source to a surface storage or settling basin; from storage to the system. In these cases, both low-lift and high-lift pumps should be monitored unless the amount of water stored is enough to supply fire demands for several hours. If storage is large enough, only the high-lift pumps (those delivering water to the system) require protection.

The supply main, if any, must be big enough to supply all demands at pressures high enough to be effective against fire or in an automatic sprinkler system. Also, reliability must be considered and special construction methods used to avoid interruption of service due to floods, landslides, or other disasters.

It is usually desirable to install an elevated water tank in an outside protection system, especially if sprinkler systems are supplied by it. An elevated tank not only provides a reliable, although limited, secondary source of water supply, but it also acts as an equalizer on the system. In other words, it assures a constant static pressure and helps to avoid serious drops in pressure during the operation of hydrants or other severe drains.

In small systems, pressure tanks will prevent drops in water pressure. These large cylindrical steel tanks are airtight, designed to withstand fairly high pressures, and installed horizontally. Such a tank is connected to the water system and to an air compressor at its top. After being filled to about two-thirds of its capacity with water from the system, compressed air is pumped in until the required pressure is reached. Pressure is maintained by an automatic control on the compressor. The disadvantages of this type of installation are relatively small storage and another threat to reliability—the compressors may break down. (For standards for both types of tanks, see "Standard for Water Tanks for Private Protection," NFPA No. 22.)

Underground Pipe

Testing laboratories, such as Underwriters' Laboratories, test and list pipe suitable for private fire protection, including such types as asbestos-cement, cast-iron, and steel. Selecting the type and class of pipe to be used in the underground system depends on soil conditions and operating pressures.

No pipe smaller than 15 cm (6.0 inches) in diameter (ID) should be used to supply hydrants; where practical, pipe lines should be "looped" so water will be supplied from at least two directions. Where looping is not used, larger pipe of not less than 20-cm (8.0-inch) diameter (ID) should be used to supply hydrants on a dead end.

How deep pipe must be buried depends upon local freezing conditions, but it must always be well below the maximum frost penetration. Since water in a fire system of this type is not normally moving, it will freeze faster than in a system supplying domestic consumption. Even in non-freezing territory, pipe should be covered by at least .76 m (2½ feet) of fill to protect it from mechanical damage; where it passes under streets, driveways, or railroad tracks, it should have a minimum cover of between .9 to 1.2 m (3.0 to 4.0 feet) depending on traffic. Special protective construction is sometimes necessary.

Improper laying of pipe causes a great deal of trouble that could easily be avoided. The trench should be smooth and pipe so laid that it is not supported by its joints but has bearing along its full length. Joints should be tight, and all tees, plugs, caps, bends, and hydrants on the pipe should be anchored. Before the pipe is covered, the system should be put under pressure and carefully inspected.

Hydrants

Hydrants should have not less than a 15-cm (6.0-inch) connection to the main and should be approved and listed for fire protection equipment by a recognized testing laboratory. There should be enough hydrants, so placed, that they offer hose stream protection for every part of each building and can concentrate required fire flow with no hose line exceeding 152 m (500 feet). Normally, hydrants should be about 15 m (50 feet) away from the building protected. This is to prevent injury by falling walls or the possibility of fire fighters being driven away by smoke and heat. Where buildings are congested and it is not possible to maintain distance, hydrants should be located with these dangers in mind and to take advantage of low buildings, masonry walls, or other protection.

Establishments that depend upon fire fighters who do not have hose-carrying vehicles must provide a hose house over each hydrant in an outside protection system. The hydrant should be as close to the front of the house as possible, but with enough room back of the doors allowed to attach the hose and operate the hydrant. Hose houses should be of substantial construction, on good foundations, painted, and well maintained. Besides the hose attached to the hydrant, spare lengths should be stored, as well as nozzles, playpipes, axe, hydrant wrenches, and spanners.

Where large amounts of combustible material are present, as in lumberyards, a fire may require large quantities of water at high pressures. Installing permanent monitor nozzles, capable of directing large quantities of water to the base of the fire and of being operated by one man, are the best answer. Locating this type of equipment and designing its piping and water supply should be done by someone experienced in fire protection.

(For details of planning, installation, and maintenance of outside protection systems, see "Outside Protection," NFPA Standard No. 24.)

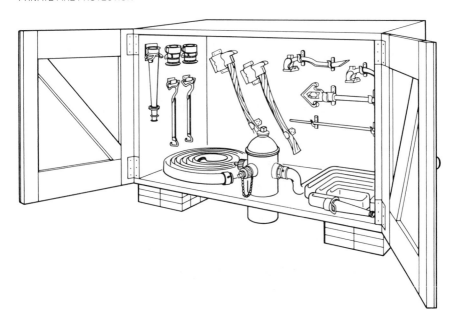

FIGURE 8-5. Hose house for outside protection system.

REVIEW QUESTIONS

1. What are the duties of a Director of Fire Safety or a Fire Loss Prevention Manager of a large manufacturing company?

2. What are the duties of the chief of a private fire brigade?

3. How are members of a private fire brigade selected?

4. How can a brigade chief stimulate cooperation with employees not in the brigade?

5. What are the functions of the guard service?

6. Describe some guard supervisory systems.

7. Describe and explain how the following first-aid fire protection appliances are used:
   (a) barrels and buckets
   (b) sand buckets
   (c) fire blankets
   (d) garden hose

8. Name the four requirements to make portable fire extinguishers effective.

9. What are the four basic types of fires?

10. Explain the following markings on fire extinguishers:
    (a) 4A : 16B : C
    (b) 1B
    (c) 2A
    (d) 1D

11. What are Class I, II, and III services when applied to standpipe and hose systems?

12. What is a "dry standpipe"?

13. Describe how water is supplied to high-rise buildings.

14. When are combined fire and domestic systems acceptable for outside protection?

15. Name two precautions that should be taken if an outside protection system depends upon a single pump for source of supply.

16. What are two advantages to having an elevated water tank in an outside protection system?

17. What is a pressure tank and how does it work?

18. What precautions should be taken in selecting and installing underground pipe for outside protection?

19. Name some standards for the design and location of fire hydrants.

20. What are permanent monitor nozzles and when is their installation justified?

9 Sprinkler and Detection Systems

Automatic Sprinkler Protection

A sprinkler system is by far the most important type of private fire protection. Knowledge of its capabilities and limitations is essential to anyone studying fire science or specializing in private fire protection work. Once again, however, we can only discuss important features, leaving details to a special study of the subject.

The recognized standard for automatic sprinkler protection, "Sprinkler Systems" (NFPA No. 13), defines a sprinkler system as:

> . . . integrated system of underground and overhead piping designed in accordance with fire protection engineering standards. The installation includes a water supply, such as a gravity tank, firepump, reservoir or pressure tank and/or connection by underground piping to a city main. The portion of the sprinkler system above ground is a network of specially sized or hydraulically designed piping installed in a building, structure, or area, generally overhead, and to which sprinklers are connected in a systematic pattern. The system includes a controlling valve and a device for actuating an alarm when the system is in operation. The system is usually activated by heat from a fire and discharges water over the fire area.*

Sprinkler systems were first introduced into the United States about 1878 in the form of a perforated pipe connected to a water supply that could be turned on in case of fire. The sprinkler head was developed a few years later. Systems have grown increasingly effective with time.

* From "Sprinkler Systems," NFPA No. 13, copyright 1975 by the National Fire Protection Association, Boston. Reprinted by permission.

Today, NFPA fire records show that more than 96 percent of all fires occurring in buildings equipped with automatic sprinklers have been satisfactorily extinguished or held in check. In many of the cases listed as "unsatisfactory," the water had been shut off or the sprinklers could not operate for some other reason.

Another important record of all fires in sprinklered buildings shows that in more than 80 percent of these cases, the fire was controlled by the operation of *seven* sprinkler heads or less; when only the more common wet pipe systems were considered, more than 80 percent of the fires were extinguished by the operation of *five* heads or less.

To work properly, all portions of a building, including concealed spaces, wide shelving, closets, cabinets, etc., must be protected by the sprinkler system. If portions are to be left without this protection, they should be cut off from the protected part by standard fire separations. This precaution not only reduces the area exposed to a single fire, but also eliminates the possibility of heat from the fire traveling along the ceiling and opening sprinklers that could not extinguish the fire. Similarly, draft stops are recommended in large-area buildings. These are curtains, built of noncombustible material such as sheet metal, that extend down from the ceiling to at least three inches below the sprinklers and reduce horizontal drafts. Vertical openings should also be enclosed to divide multistoried buildings into separate fire areas.

Water Supplies

Every automatic sprinkler system must have at least one water supply with adequate pressure and volume to meet its demands. This supply must also be reliable and be available without human assistance when the system demands. A secondary supply is sometimes required, depending upon a variety of conditions, including degree of reliability of the primary supply, hazard of the occupancy, number of sprinklers that may operate, rate of discharge needed from each sprinkler, required time of discharge, and amount of water needed for hose streams.

When we discuss fire flow requirements for municipalities (Chapter 10) as for outside protection, we shall see that although there are guides, there is no rule or formula to fix the amount of water needed for fire protection. The same is true for sprinkler supplies.

The standard* provides a guide, but this must be interpreted in the light of experience. For light hazard occupancies, the amount of water required varies from 1140 lpm (300 gpm) for a small system up to 2840 lpm (750 gpm) for a large installation. Ordinary hazards require from

---

* For description of light, ordinary, and extra hazardous occupancies, see Appendix C.

2650 lpm (700 gpm) to 5678 lpm (1500 gpm), or even more. In applying this guide, and in all extra hazard occupancies, the design must take into account combustibility of contents, areas shielded from proper distribution of water, height of stock piles, type of ceiling construction, ceiling heights, undesirable draft conditions and size of undivided areas. After the amount of water a given sprinkler system will need has been fixed, the supply must be tested to see if the water system has enough pressure to deliver this amount of water to the highest line of sprinklers with at least 103 kPa (15.0 psi) residual pressure.

Sources of supply for sprinkler systems are about the same as those we discussed under "Outside Protection," but there are additional specific requirements. For instance, an elevated tank capacity should be large enough to supply the required water for 60 minutes in a light hazard occupancy and for 60 to 120 minutes in an ordinary hazard occupancy. Except in unusual cases, pressure tanks are acceptable only if there is also a reliable secondary supply and if the pressure tanks hold at least 7570 liters (2000 gallons) for light hazard occupancies and 11360 liters (3000 gallons) for ordinary hazard occupancy.

Another sprinkler system source of supply is the fire department connection. This is usually at least a four-inch pipe connected to the system through a check valve and extending through the wall of a building to the street side. Two or more hose connections allow the fire department to pump water into the system from an outside source.

System Components

When installing piping for sprinkler systems, the builder should use only specific types, tested and approved for this purpose. Usually, these types are of wrought steel or wrought iron, although other types of pipe or tube may be found where conditions require it. Regardless of type, however, all sprinkler pipe must be able to withstand a working pressure of not less than 1200 kPa (175 psi). Figure 9-1 shows the details of an average sprinkler system's piping. The NFPA standard has schedules of pipe sizes, based on the number of sprinklers in the system, for light, ordinary, and extra hazard occupancies. Branch lines must not be smaller than 2.5 cm (1.0 inch) and cross mains must not be less than 3.2 cm (1¼ inch) at any point. Risers should be big enough to supply all the sprinklers fed by them on any one floor of a fire section. Frequently, systems are designed hydraulically, rather than using pipe schedules. This is particularly true of large systems. This method, based upon the hydraulic computation of friction loss in pipes and using the sizes needed to deliver required quantities, results in a better balanced system and often reduces cost.

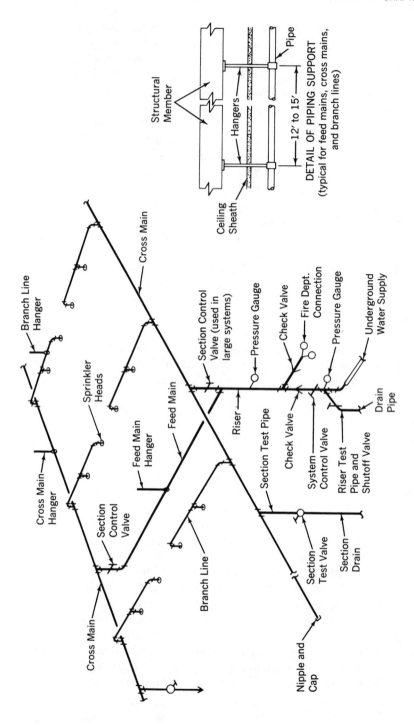

Structural Member

Hangers

Pipe

12' to 15'

DETAIL OF PIPING SUPPORT
(typical for feed mains, cross mains, and branch lines)

Ceiling Sheath

Cross Main

Branch Line Hanger

Sprinkler Heads

Section Control Valve (used in large systems)

Pressure Gauge

Check Valve

Fire Dept. Connection

Pressure Gauge

Underground Water Supply

Feed Main Hanger

Feed Main

Riser

Section Test Pipe

Check Valve

System Control Valve

Riser Test Pipe and Shutoff Valve

Drain Pipe

Cross Main Hanger

Cross Main

Section Control Valve

Branch Line

Section Test Valve

Section Drain

Nipple and Cap

FIGURE 9-1. A typical sprinkler system.

When a system is hydraulically designed it is necessary to determine the required density (lpm per square meter or gpm per square foot) distributed with a reasonable degree of uniformity over a specified area. These densities are shown in the standard and range from 0.024 lpm per m² (0.070 gpm per sq. ft.) for 278.7 m² (3000 sq. ft.) of light occupancy to 0.088 lpm/m² (0.250 gpm per sq. ft.) for 139.4 m² (1500 sq. ft.) of ordinary group 3 occupancy. For higher hazard occupancies special design criteria are used.

To allow for flushing the sprinkler system, there should be a 5-cm (2-inch) capped nipple at the end of each cross main and all pipes should be installed with the proper slope or pitch to permit thorough draining of the system. Usually, this is done by sloping the pipe so it will drain toward the source of supply and by putting a drain valve at the base of the riser. Where sectional control valves are used, a drain valve is put in at a point where it can completely drain the portion of the system controlled by the sectional cutoff valve.

Well built systems also have a test pipe, at least 2.5 cm (1.0 inch) in diameter, attached to the top of the system at some convenient point (either to a cross main, branch line, or the top of the riser). Flow from the test pipe should be equivalent to that of a sprinkler and the arrangement should allow an inspector to see the flow of water while operating the valve on the pipe. Figure 9-2 is an example of a certificate issued after testing.

In cold weather areas, where supply pipes and risers pass through unheated portions of buildings, the sprinkler system must be protected against freezing. Where corrosive conditions exist, all piping and fittings must be of corrosion-resistant material with a protective coating. In areas subject to earthquakes, piping needs special installation to protect it against breakage from building or land movement.

Control valves are perhaps more important in a sprinkler system than in any other type of water system. All sources of water supply, except fire department connections, should be controlled by a gate valve that shows at a glance whether it is open or closed. There should be a check valve on all supply lines to let water enter the system but prevent flow in the other direction. In buildings of large area, excessive height, or for other reasons, it may be desirable to install floor or sectional control valves. These should be an indicator type valve in the feed main at a readily accessible point.

Piping should have substantial support from the building structure. Except for risers, support usually comes from approved types of hangers attached to structural members, not ceiling sheaths. For steel and wrought iron pipe, distance between hangers should not exceed 3.7 to 4.6 m (12 to 15 feet), depending upon pipe size. Copper tube requires

FIRE DEPARTMENT

FIRE PROTECTION EQUIPMENT PERFORMANCE CERTIFICATE

LOCATION AND TYPE OF EQUIPMENT TESTED	
DBA:	
ADDRESS:	TELEPHONE:
TYPE OF SYSTEM TESTED:	

DEFECTS FOUND:	
REPAIRS MADE:	
INITIAL TEST DATE:	FINAL TEST DATE:
SERVICING FIRM:	ADDRESS:
SIGNATURE:	TITLE:

I hereby certify that the fire protection equipment indicated above has been tested in accordance with Regulation Number 4 of the Fire Department. All necessary maintenance and repair have been made in compliance with the Municipal Code. To the best of my knowledge the equipment tested is operable.

Building Owner or his Agent.

Title: _____   Signature _____

Date: _____

FIGURE 9-2. Example of fire protection equipment performance certificate.

closer spacing. Risers are supported by being attached directly to the structure and at the ground.

The standard sprinkler head is the result of a complete redesign of the deflectors during 1952 and 1953. This design creates a spray pattern shaped much like an umbrella with the open side down. (See Figure 9-3.) The discharge covers a circular area; very little water is discharged upward to wet the ceiling. An average sprinkler has a 1.3-cm (0.5-inch) discharge orifice and can deliver between 76 and 230 lpm (20 and 60 gallons per minute) at pressures from 69 to 689 kPa (10 to 100 psi). Sprinklers should *never* be painted or treated in any way since this may change the characteristics and temperature at which they operate. Where corrosive conditions exist, specially coated approved sprinklers can be obtained from the manufacturer.

An ordinary sprinkler will operate at temperatures between 54.4 and 73.9°C. (130° and 165°F.). This type is used where *ceiling* temperature does not exceed 37.8°C. (100°F.). Intermediate sprinklers are used where ceiling temperatures do not exceed 65.6°C. (150°F.) and operate between 79.4°C. (175°F.) and 100°C. (212°F.). High, extra-high and very extra-high rated heads may be obtained for unusual ceiling temperatures up to 246°C. (475°F.); there are even some available at 316°C. (600°F.). Depending on the size of the system, there should be a stock of from 6 to 24 sprinklers on the premises for replacement purposes.

Every sprinkler system should be equipped with a water flow alarm. This apparatus sounds an alarm whenever there is a flow of water in the system equal to or more than that from a single sprinkler. The best practice is to have this device connected to a central alarm station or the fire department.

Pressure gauges are required at or near the inspector's test pipe and in sprinkler risers, above and below each alarm check valve.

Location and Spacing of Sprinklers

Fire protection engineers who follow the NFPA standard can be certain that any sprinkler system they design will perform properly and give the best protection possible. The standard lists areas that might normally be overlooked, such as basements, lofts, concealed spaces, under decks, under platforms and tables. It also gives the maximum area of protection afforded by each sprinkler, depending upon type of building construction and the hazard rating of the occupancy. The standard also describes where to place sprinklers so that there will be minimal interference with discharge from beams, girders, air ducts, and other structural features and no decrease in their sensitivity to the heat that makes the system operate. The next three paragraphs illustrate how carefully NFPA standards are prepared.

Upright   Pendent

a. Standard (modern) automatic sprinklers

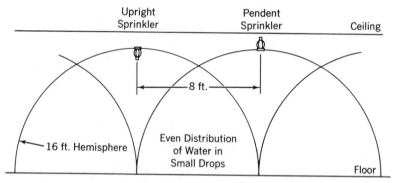

Upright   Pendent
Sprinkler   Sprinkler   Ceiling

◄——— 8 ft. ———►

16 ft. Hemisphere   Even Distribution
of Water in
Small Drops
                                                Floor

b. Typical modern sprinkler pattern

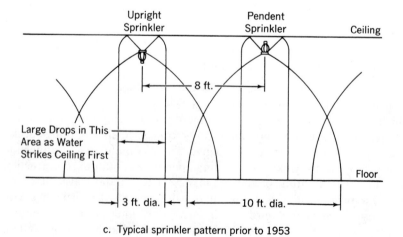

Upright   Pendent
Sprinkler   Sprinkler   Ceiling

◄——— 8 ft. ———►

Large Drops in This
Area as Water
Strikes Ceiling First

                                                Floor

►| 3 ft. dia. |◄—— |◄———— 10 ft. dia. ————►|

c. Typical sprinkler pattern prior to 1953

FIGURE 9-3. Comparison of sprinkler patterns before and after 1953.

In light hazard occupancy, maximum allowable distance between branch lines and between each sprinkler on a branch line is 4.6 m (15 feet). Sprinklers need not be staggered unless they are installed under wood joists and the distance between branch lines and sprinklers exceeds 3.7 m (12 feet). In ordinary hazard occupancies, the same limitations apply, except where there is high-piled storage; in this case, the distance between each sprinkler and the branch lines themselves is reduced to 3.7 m (12 feet). In extra hazard occupancies, the distance between sprinklers and branch lines is 3.7 m (12 feet), and sprinklers should be staggered in all cases.

Average protection area per sprinkler varies from 18.6 m² (200 square feet) in light hazard occupancies with smooth ceiling construction, to 8.4 m² (90 square feet) in extra hazard occupancies of any type construction.

In certain locations, such as show cases or spaces where head room is limited, installing conventional overhead sprinklers is impractical. Sidewall sprinklers may be substituted for use in light or ordinary hazard occupancies. These installations will meet space requirements and protection area limits. Installing sidewall sprinklers calls for expert advice. They must be so placed that they will receive heat from a fire as quickly as possible and distribute water effectively. Only a special type of sprinkler, approved for this service, should be used and the standard's advice about the contour of the space must be followed.

Types of Sprinkler Systems

By far the most common sprinkler system is the *wet pipe* type. This is used where piping is not subject to freezing, either in parts of the country where there are rarely cold temperatures or in heated buildings. A wet pipe sprinkler system consists of automatic sprinklers attached to piping that holds water and is connected to a water supply. When heat melts the sprinkler's fusible element, water discharges immediately, extinguishing any fire in the area.

The next most common form of sprinkler protection, the *dry pipe* type, is used in rooms or buildings that cannot be properly heated. This type consists of sprinklers attached to piping that contains air under pressure. When air is released by actuation of a sprinkler, this opens a dry pipe valve; water flows into the system and is discharged from the opened sprinkler. When 25 percent (or less) of the sprinklers need to be on a dry pipe system, this portion of the entire system frequently draws its water from a wet pipe system, again through a pressure valve. Since the dry pipe valve, in either a full or partial dry pipe system, is kept from opening by a higher air pressure than the pressure of its water supply, the source of compressed air must be reliable. Gauges are supplied on both the air and water side of the dry pipe valve and air pressure maintained at 103 to 138 kPa (15.0 to 20.0 psi) in excess of tripping pressure.

A *deluge system* consists of *open* sprinklers attached to piping of comparatively larger size than in other types of systems. The control valve to the water supply is operated manually or by a supplementary system of heat responsive devices. The heat responsive device covers the same area as the sprinkler system and spacing of the system's parts is determined by laboratory test. *Deluge systems* are used where conditions of occupancy or special hazards require quick application of large quantities of water, such as in airplane hangars.

A *pre-action sprinkler system* is similar to a deluge system except that ordinary closed sprinklers are used and piping may be no larger than the standard for occupancies of the same type. The difference between *pre-action* and dry pipe systems is that the control valve is operated by heat actuated devices and not by the opening of a sprinkler. Since heat actuated devices are more sensitive than sprinklers, the control valve opens sooner and a local alarm, operated by the control valve, gives quicker warning of fire. Pre-action systems are of particular value in protecting stocks of material that are particularly subject to water damage.

The pre-action system can be combined with the dry pipe system when two or more dry pipe valves are needed to supply the system. In this combination, all the piping holds air under pressure; the area is also covered by a heat actuated device that operates the dry pipe valves simultaneously. Water is admitted, pushing the air through exhaust valves at the end of the system, and making water immediately available when the sprinkler head opens. This type of system is used to cover long piers in cold climates where protecting water supply from freezing is impractical.

A recent development is the *recycling type sprinkler system*. This is a conventional wet pipe system that supplies water when a sprinkler is open, but that automatically closes the main valve when the fire is out. Shutoff is controlled by a supplemental heat actuated device that covers the area protected by the sprinkler system and actuates solenoid switches that operate the control valve. When heat is reduced by fire extinguishment, the supply valve closes; should the fire rekindle, the heat actuated device again opens the valve. This type of system has unlimited application, but its principal disadvantage is cost.

Outside Sprinklers (Water Curtain)

Normally, sprinkler systems inside buildings control a fire from any source effectively, usually with the operation of only a few sprinklers. But under certain conditions, radiated heat from a fire *outside* a sprinklered building can open enough sprinklers to weaken the system and make it ineffective. To prevent this, seriously exposed buildings with unprotected openings or of combustible construction are equipped with an outside sprinkler system: open sprinklers, specially designed to direct water against the wall or windows and underneath the cornice of the building. Frequently, these outside systems are manually operated. The source of water should be able to supply the entire outside sprinkler system for at least one hour while maintaining pressure of 69 kPa (10 psi) at the top of the riser. There should also be a separate fire department connection. Manual operation has unreliable features; chiefly, someone must always be on the premises to operate it. Automatic operation is preferable with the outside sprinkler system designed to operate like a deluge or pre-action type. Where there is enough water available to supply an outside system without seriously depleting the inside supply, it is sometimes desirable to feed the outside system through a dry pipe valve from inside the building.

Piping should be galvanized or otherwise protected from corrosion and installed to permit thorough draining. Not more than 40 sprinklers are permitted under one control and drain valve. This is to conserve water. Small orifice sprinklers are used when the building is not severely exposed, but if the exposure is serious and there is an adequate supply of water, large orifice sprinklers are used, especially on upper branch lines.

Sprinklers outside of a building need not be as closely spaced as overhead inside sprinklers. Water from the top line of outside sprinklers will flow down, wetting the walls and windows below. On a three story building, only one line of sprinklers along the top is required, unless windows are deep-set or overhanging sills prevent thorough wetting. Under these conditions, a window sprinkler is needed at each story but these sprinklers may have smaller orifices.

In mild climates where freezing is not a problem, exposed openings are frequently protected by simply extending a branch line of an interior wet pipe system through the wall with a sprinkler attached.

Other Automatic Systems

As we have seen in our discussion of extinguishing agents in Chapter 6, water is not always the most effective means of controlling a fire. This applies equally to sprinkler systems. They are usually the best type of private fire protection, but there are some special hazards that can be handled more effectively by another type of system.

Carbon Dioxide Extinguishing Systems

Besides being a very effective agent for extinguishing flammable liquid and gas fires, carbon dioxide has the advantage of leaving no residue;

it also has no adverse effect of any kind on the burning material. For these reasons, carbon dioxide systems are used to protect diptanks, spray painting booths, and flammable liquid hazards where carbon dioxide will extinguish the fire before any metal in the area can be heated to the point where it might rekindle the flammable liquid vapors. They are also recommended to protect small areas where the contents are especially susceptible to water damage, such as fur and record vaults and computer rooms and equipment. Theoretically, there is no limit to the space that can be protected with carbon dioxide; practically, the quantity of gas required to create a smothering atmosphere restricts its use to relatively small enclosures. Occupants of an area protected by a carbon dioxide system must have ample warning to allow them time to vacate the area before the gas is discharged.

NFPA Standard No. 12 covers four types of carbon dioxide extinguishing systems:

- *Total flooding.* This is a supply of carbon dioxide connected to fixed piping with nozzles arranged to discharge the gas into an enclosed space or enclosure about the hazard.
- *Local application.* A supply connected to fixed piping with nozzles arranged to discharge directly on the burning material.
- *Hand hose line.* A supply of carbon dioxide connected to hose lines.
- *Standpipe, with mobile supply.* This has a source of carbon dioxide that can be quickly moved and connected to a system of fixed piping, supplying either fixed nozzles or nozzles on hose lines for flooding or local application.

Carbon dioxide is stored in two ways: under high or low pressure. The system is referred to accordingly. The high pressure system is more common and is supplied by a number of cylinders containing liquid carbon dioxide at a pressure of about 5860 kPa (850 psi) at temperatures of 21°C. (70°F.). These cylinders are designed to meet Department of Transportation specifications and will safely withstand pressures up to 20 670 kPa (3000 psi). Low pressure systems are supplied by pressure vessels designed to withstand 2240 kPa (325 psi) with the carbon dioxide maintained at about -18°C. (0°F.) by refrigeration and insulation around the container. This type of storage is especially useful when large quantities of gas are needed, as these vessels can be made large enough to store up to 112 500 kg *(125 tons)* of carbon dioxide.

Total flooding and local application systems are usually designed to operate automatically but all automatic systems are required to have an independent means of manual operation. A system of detectors, sensitive to heat, smoke, flame, or other unusual conditions, actuates automatic operation. A detector may actuate a number of processes: giving an alarm, opening a valve that releases carbon dioxide into the piping system and out the open nozzles in the area to be protected, shutting off ventilating fans, closing doors and windows, or other necessary functions. In what are known as "extended discharge" systems, after initial discharge, the carbon dioxide is shut off for a predetermined period; a secondary discharge follows which may be continuous or intermittent. This is valuable where smoldering material may remain after the first application, threatening to rekindle.

Carbon dioxide systems, incorporating hand hose lines or standpipes with mobile supplies, depend on manual operation by trained personnel. They are not a substitute for fixed systems, but can be used to supplement both fixed systems and portable equipment.

Foam Systems

As we saw in Chapter 6, although several extinguishing agents have been approved for flammable liquid fires, only foam has been found practical for combating fires in large, outdoor storage tanks. Foam extinguishing systems are frequently used in large oil refineries, tank farms, or wherever large quantities of flammable liquids are stored.

Foam systems, to protect against localized inside hazards, can be operated automatically by heat actuated devices in much the same way as carbon dioxide systems. But, because of the complication of foam generation and the space required to store the materials, other automatic systems are generally preferred. Standards for foam extinguishing systems are detailed in NFPA Standard No. 11.

In large structures with flammable liquid hazards, such as aircraft hangars and some chemical processing buildings, sprinkler systems combine foam application with water. *Foam-water systems* (see NFPA Standard No. 16) consist primarily of a deluge-type sprinkler system with specially designed open sprinklers that also serve as foam nozzles. The system puts out mechanical foam for a predetermined period, and then operates as a deluge system to extinguish any smoldering Class A material that might rekindle.

High expansion foam, described in Chapter 6, can also be applied through an automatically operated system. See NFPA Standard No. 11A, *High Expansion Foam Systems.*

There are three types of high expansion foam systems: total flooding systems, local application systems, and portable foam generating devices. The first two types, like other automatic foam systems, are operated by a system of heat actuated devices that control foam generators, valves, proportioners, eductors, discharge controls, and shut-down equipment. These systems may flood an entire building, yet the amount of water that will be present is so small that no appreciable damage results. Air in the foam is sufficient to sustain life so that fire fighters can walk through a foamfilled building.

Chemical foam systems have been almost completely replaced by air foam systems and no new chemical foam systems are being installed.

The "combined agent" or "twinned" system has not been automated, because judgment is required to obtain the desired results. This system consists of two nozzles each with trigger control fastened together in such a way that one person can operate both. One nozzle is supplied through a hose line with aqueous film-forming foam (AFFF) and the other with dry powder. The dry powder is applied first to extinguish the fire, followed immediately by the application of AFFF to seal off the vapors and prevent reignition. The system has been adapted to mobile devices for one-person operation and is especially valuable in fighting aircraft fires. For details, see NFPA Standard 11B, *Synthetic Foam and Combined Agent Systems.*

Halogenated Extinguishing Agent Systems

The history of halogenated extinguishing agents has been one of peaks and valleys. The word *halogenated* means to be treated with or caused to combine with a halogen. A halogen is one of the following five elements: fluorine, chlorine, astatine, bromine, or iodine. There are a great many halogenated compounds, but only a comparative few have been used for fire extinguishment.

The first one appeared in the early part of this century in the form of carbon tetrachloride (Halon 104). It was spectacular because of its ability to put out Class C fires and even small Class B fires. The extinguishers were of small size and called "vaporizing liquid" extinguishers. In the 1920s this type of extinguisher reached its peak. We discussed the demise of the carbon tetrachloride extinguisher, however, because of its toxic and corrosive properties, in Chapter 8.

Similar ups and downs were experienced with other halons until 1962 when Halon 1301 was approved by nationally recognized laboratories. This was no longer a vaporizing liquid, since Halon 1301 has a boiling point below 0°C. (freezing), and these extinguishers are sometimes referred to as liquefied gas extinguishers.

Used in hand extinguishers for many years, Halon 1301 has been adapted to automatic extinguishing systems and NFPA technical committees have prepared standards for systems using two of the halons. See Standards No. 12A, *Halogenated Extinguishing Agent Systems— Halon 1301*, and No. 12B, *Halogenated Extinguishing Agent Systems— Halon 1211*. These systems discharge dry chemical forced through piping or hose lines by compressed gas from fixed nozzles. Since dry particles moving through pipelines do not follow hydraulic principles and since tests have shown that powders differ in flow characteristics, each system must be designed by and installed under the supervision of a fire protection engineer who is knowledgeable in this field.

Halon systems are particularly effective on flammable liquids and gases, combustible solids that melt when involved in fire (like pitch or tar), electrical hazards, ordinary combustibles, and kitchen hoods, ducts, and so forth. They are not recommended for use on fuels containing their own oxidizing agents (such as gunpowder, rocket propellants, cellulose nitrate, etc.), reactive metals (such as sodium, potassium, magnesium, titanium, etc.), deep-seated or burrowing fires (such as in cotton bales), or where the residue might affect electronic or other delicate equipment.

Like carbon dioxide, these systems may be total flooding, local application, or hand hose line type. The first two may be automatic or manual, but if automatic they must have provision for manual operation.

The halogenated agent is usually contained in one or more pressurized vessels near the protected area. Steel cylinders or spheres holding from a few kilograms to several hundred kilograms, depending upon the area to be protected, are in use. Superpressurization of the storage containers is required by NFPA standards for Halon 1301 and Halon 1211. This extra pressure is necessary to completely expel the agent at all normal temperatures and provide a uniform extinguishing concentration throughout the area. In a properly designed system the agent is capable of extinguishing a fire regardless of its location within the area or enclosure.

Dry Chemical Extinguishing Systems

A review of Chapter 6 will remind the reader that dry chemical extinguishing agents usually have a potassium or carbonate of potassium chloride base and are exceptionally rapid extinguishers of flammable liquid fires. The multipurpose dry chemical agents are also effective on Class A fires and, since they can be safely used on most electrical equipment, they are listed for use on Class A, B, and C fires.

This agent is adapted to fixed-system use very much the same way as are halogenated agents. There is usually a storage of dry chemical and the expellent gas under pressure. The gas is usually nitrogen, but carbon dioxide is sometimes used in small installations. The system can be actuated either by heat detectors or manually and it can be adapted to the use of hose lines.

Fixed Water Spray Systems

In Chapter 6, we discussed the use of spray or fog by fire departments. Fixed water spray fire protection systems depend on the same principles

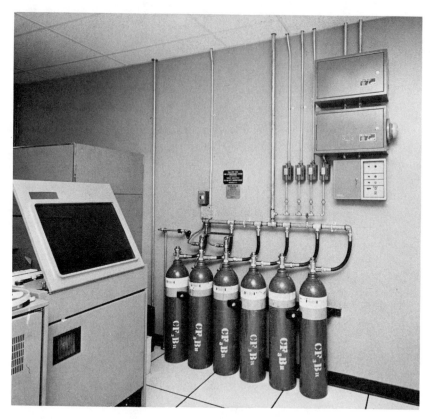

FIGURE 9-4. Halon extinguishing gas system interfaces with smoke detectors. In addition to initiating a flow of gas, the fire alarm system also signals a central station. Courtesy of the American District Telegraph Company.

to protect against specific hazards. They are most effective on electric transformers, oil switches or motors, ordinary combustibles, and certain hazardous solids. Adding wetting agents that increase water penetration of some solids and provide foaming characteristics extends the use of water spray.

The fixed water spray system is a sprinkler system with specially designed sprinklers or nozzles that give the most effective *spray pattern* against a particular hazard. The system may be independent, or it may supplement other forms of protection. Most systems of this type operate with open nozzles, resembling the deluge type. Spray pattern and discharge rate are based upon experience and tests made under controlled conditions. Since many types of nozzle have restricted waterways, there must be a removable strainer in the supply line to screen out anything that would block or interfere with the flow of water. These strainers must be cleaned and flushed at regular intervals. Some types of nozzle require individual strainers; these also need careful maintenance.

Water supply requirements are much the same as for sprinkler systems, including a fire department connection. Where water supply is extremely limited, the NFPA standard provides for a cycle system. Water discharged by the system is collected in a fire drainage trench and directed into an interceptor system that separates foreign material. The water is then pumped into the system and used again. Care must be taken that flammable liquids are not introduced into the water spray system.

A self-contained water spray system has been developed to protect an isolated hazard such as a transformer bank or a tank of valuable flammable liquid where water supply is extremely limited or nonexistent. A heat detector over the hazard opens a valve; nitrogen under pressure enters the water tank and forces the water through the system's piping to water spray nozzles over the hazard. The system can also be activated manually. Tests indicate that, when protected by a water spray system, most fires in flammable liquid containers can be extinguished in 10 to 45 seconds. The size of the water tank is based upon anticipated extinguishing time plus a safety factor. The nitrogen supply should be sufficient to expel the entire water supply in the tank.

Just as for other widely used fire protection systems, standards for fixed water spray systems are published by the National Fire Protection Association (NFPA No. 15) and cover details of water supplies, piping, valves, alarms, and so on.

Protective Signaling Systems

Fire protection includes life safety, fire suppression and control, fire prevention, fire investigation, and fire detection. So far, we have said very little about fire detection, other than what was covered in our discussion of sprinkler systems. Protective signaling systems, however, play a major role in fire detection. This is because they can be adapted to many purposes, such as:

- Notifying people of a fire so that they can leave the premises.
- Calling the fire department or private fire brigade.
- Monitoring extinguishing systems.
- Monitoring industrial processes and warning of hazardous conditions.
- Supervising people, such as guards, electronically.
- Actuating control equipment.

Of these six possible functions, three, the first two and the last, depend upon accurate and rapid detection of a fire by the system; the other three may fairly be classed as fire prevention measures since they deal with conditions that may result in a fire.

Alarms

An alarm may be as simple as a hand-operated gong, or it may depend upon a complex electrical system that can be operated from many different places or that acts on electronic command such as a break in power service. Regardless of what kind of alarm system is used, to be effective it must observe five principles:

- The alarm must be distinctive; all occupants should recognize it as a fire alarm at once. There is a major exception to this rule:

  In hospitals, jails, and other institutions where an alarm could cause a panic because the inmates cannot leave, those responsible for evacuating people should get the signal.

- All occupants must receive the warning. In very noisy locations or where deaf people work, there should be a visual signal as well as a sound signal.

- The system should be arranged so that it can be tested periodically.

- The system's design and installation should make it difficult, if not impossible, to operate by accident, as a joke, or maliciously.

- Especially in places of public assembly, there must be a choice between having the alarm go to all occupants or to a selected group of people only.

Because the first few minutes count very heavily in fire suppression, the second function of a protective signaling system, calling organized assistance, ranks high. The best system is one that warns people and summons help simultaneously; if this is not possible, assistance should be called at once. The established hotel policy of investigating a fire before calling the fire department in order to avoid disturbing guests has led to several major disasters with heavy loss of life. The fire department would much rather respond to a call and find it is not needed than to arrive late and find a fire that is out of control. Whether a system or person calls for assistance, a protective signal system should clearly indicate the location of a fire or other dangerous conditions by building, and preferably by floor, area, or even room.

Effective monitoring of extinguishing systems calls for detection devices or systems that will sense or react to anything that will interfere with or prevent operation of the extinguishers and, if such a condition exists, give a distinctive warning signal, often called a trouble signal. Since it is possible to have a trouble signal triggered by almost anything that can be measured or detected, there are many different ways to monitor extinguishing systems. In sprinkler systems, for example, we can monitor air pressure in dry pipe systems; water level in tanks and reservoirs; condition and performance of valves that control operation of the system; electrical power supply to pumps, and even the operation of the pumps themselves. The same principles can be applied to monitoring heating systems, air conditioning systems, and transformers.

In Chapter 8, under *Guard Service,* we discussed several ways of controlling how people perform their duties, as well as the supervisory function of those in charge. A protection system adds to the value of guard service when it combines patrol stations with manual fire alarm boxes. Such an arrangement lets the guard call for assistance quickly.

Classification of Systems

Protective signaling systems are classified according to the location of the signal-receiving equipment. Local alarm systems are those producing signals at one or more places on the premises served. A simple and common example of this classification is the water flow alarm on a sprinkler system. This may be a gong, mechanically operated by a water motor in the supply line, or an electrically operated bell powered by a battery or a connection to the building's electrical power supply on the line side of the main service. A more complex type of local system is frequently found in apartment buildings or similar occupancies. Here, the system operates on a closed circuit principle: when power supply is interrupted, a second power source, usually a battery, sounds the alarm. Alarm boxes or stations are essentially switches on the circuit, connected in series; any one may open the circuit and send in an alarm. There are many variations of the local system, but its three essentials are: a reliable source of power, proper wiring material and methods, and a monitoring circuit that will give a distinctive signal in case of trouble and faithfully transmit fire alarms.

Remote station systems use electrically monitored circuits running directly between the signaling devices at the protected premises and signal-receiving equipment in a remote station, such as a municipal fire alarm headquarters, a fire station, or a private agency equipped and staffed to receive the signals and act upon them. This type of system is frequently used to protect premises that are not constantly guarded. It is used more frequently with automatic alarm systems than with manual. Systems of this type are generally leased by the owner of the protected premises and maintained by the leasing company. When the remote station is a fire station, the firemen must know whom to call in case of a trouble signal; of course, the department responds to fire alarms.

Auxiliary systems are those that send signals to the fire department over municipal fire alarm circuits. Some of these systems consist of a number of local boxes on the premises that can operate a street box in the municipal system through a "shunt" circuit. This means that the system depends upon power from the municipal system for operation; if not properly maintained, it may send false signals. Another type of auxiliary system uses its own power supply and is connected to a specially designed fire alarm box, called a master box. This not only actuates the box but designates the auxiliary box or general location of the fire as well. These are sometimes operated by automatic systems and, when serviced under contract with a leasing company, have a good record for reliability.

A *proprietary system* is one supervised by competent and experienced people from a central supervising station at the property protected. The system permits the operators to test and operate it, and upon receipt of a signal, to take the action required. This type of system is usually found in large establishments. It can be thought of as an expanded local system where all signals are received at one central point. Guard service and patrol supervision are usually centered at this station, too, so it is always manned. Proprietary systems are generally owned and controlled by the plant. Management can establish the degree of reliability to a great extent by carefully selecting supervisory and maintenance personnel and outlining proper rules of procedure.

A *central station system* is defined in the NFPA Standard as:

> A system or group of systems, the operations of which are signaled to, recorded in, maintained and supervised from an approved central station, in which there are competent and experienced observers and operators in attendance at all times whose duty it shall be, upon receipt of a signal, to take such action as shall be required under the rules established for their guidance. Such systems are independently owned, controlled and operated by a person, firm or corporation whose principal business is the furnishing and maintaining of supervised protective signaling service and who have no interest in the protected properties.*

In other words, this type differs from a proprietary system in being owned and operated by a company other than the owner of the protected property. These companies own the systems, maintain them on a contract

*From *Central Station Protection Signaling Systems,* NFPA No. 71, copyright 1974, National Fire Protection Association, Boston. Reprinted by permission.

basis, and serve a great many plants in a given geographical area, usually about 518 km² (200 square miles). They maintain direct lines from their central station to the police and fire departments; messengers or runners are on duty who can be quickly dispatched if the signal indicates investigation is needed. While this type of system is generally regarded as the most reliable, it is not available except in large population centers.

Household fire warning systems are one of the three provisions necessary to provide reasonable life safety in the home, where a large percentage of the loss-of-life fires occur. The other two provisions are minimizing hazards, and having and practicing an escape plan.

This class of system, to be effective, must respond to rapidly developing, high heat fires and slow smoldering fires either of which can produce smoke and toxic gases. Statistics show that most casualties of household fires occur at night and are due to the inhalation of smoke and gas rather than burns, and most of these deaths are the result of night fires when the occupants are asleep.

Since heat detectors and sprinklers often do not operate until after the smoke and gas have done their damage, household warning systems are usually actuated by smoke detectors. By 1970 there were several effective devices of this kind on the market but they were too expensive for the ordinary household. By 1975, however, there were at least a dozen different models on the market priced low enough so the average homeowner could afford this protection. Several cities and a few states now require smoke detectors in every new dwelling.

Manual operation of a protective signaling system generally consists of using a fire alarm box. When a coded signal is not required, these devices are, in effect, simple switches. If coded signals are needed, operating the box starts a mechanical or electrical device that turns a character wheel. This wheel opens and closes the circuit at predetermined intervals, sending a coded signal. Some way to operate the system manually is recommended for automatic systems.

Automatic operation of a protective signaling system depends upon detectors: devices sensitive to heat, smoke, flame, pressure, flow of current—the list is endless. We will consider the operating principles of a few kinds a little later in this chapter. Water flow alarms on sprinkler systems are, in fact, automatic fire alarms, but this term is usually restricted to a system of fire detectors.

Standards for various types of protective signaling systems, published by the National Fire Protection Association, are as follows:

Central Station Protection Signaling Systems, NFPA No. 71
Local Protective Signaling Systems, NFPA 72A
Auxiliary Protective Signaling Systems, NFPA No. 72B

Remote Station Protective Signaling Systems, NFPA No. 72C
Proprietary Protective Signaling Systems, NFPA No. 72D
Municipal Fire Alarm Systems, NFPA No. 73
Household Fire Warning Systems, NFPA No. 74

Operating Principles of Fire Detectors

Heat detectors are of two general types: those that operate at a predetermined temperature, called fixed temperature devices, and those that operate when there is an unusual increase in temperature, designated as rate-of-rise types. Some devices combine both features. Heat detectors may also be divided into two groups according to pattern used to cover the protected area. The spot pattern type consists of heat sensitive elements, in small units at intervals over the area; the line pattern type is a continuous line or circuit.

The most common fixed temperature device is probably the fusible link: Two pieces of metal held together with solder of a known melting temperature. This is used to open sprinklers, operate fire doors, and perform various other functions but is seldom used in fire alarm systems. Another fixed temperature device, frequently used in sprinkler systems, is the quartz bulb type. This depends upon the expansion of a liquid under heat to break the containing bulb, opening the sprinkler. Both fusible link and quartz bulb devices must be replaced after any use.

The fixed temperature detector used most widely in protective signaling systems employs the principle of different coefficients of expansion in metals. By arranging electrical contacts so that a circuit opens or closes according to the difference in length of two strips of different metals, an alarm will sound at a predetermined temperature. This is the same principle as an ordinary thermostat. Another device of this type depends upon heat expanding a metal disc. The disc is so held that when it expands to a predetermined point, the center will bulge upward and make an electrical contact. Either of these thermostat types returns to the open position when cooled and does not need replacement.

A fixed temperature device, using the line pattern type, is the thermostatic cable. This consists of two steel conductors, separated from each other by a heat sensitive plastic insulation, wrapped in a braided coating to protect the assembly from mechanical injury. At a certain temperature the plastic melts, allowing the two conductors to make contact.

Rate-of-rise detectors can be set to operate more readily than fixed temperature detectors and to adjust themselves to changes in normal temperature without giving a false alarm. This permits their use in normally warm locations, such as furnace rooms, or in cold places, such as cold storage rooms. These systems also readjust themselves after operation. Pneumatic tube detectors are the most widely used type in fire alarm systems. In line pattern design, pneumatic tube detectors consist of a circuit of small diameter tubing covering the protected area. (The tubing can also be coiled to form a rosette of relatively small area for use as a spot detector where desirable.) Tubing circuits end at a transmitter having an air chamber with constant air pressure automatically maintained by an adjustable air vent. When heat expands the air in the tubing, the increased pressure is transmitted to the air chamber and the expanding air moves a diaphragm that makes an electrical contact. This contact sets off an alarm. Another type of rate-of-rise detector uses the principle of the thermocouple: when two different metals are connected in a loop and one joint is heated more than the other, an electric current is generated in the loop. In this type of detector, two or more thermocouples are so mounted that one joint is exposed to the air and the other is shielded. These detectors are connected in series with a galvanometer relay, and batteries supply a monitoring current. The relay is balanced to the monitoring current, but when an additional current is created by heating one end of the thermocouples, the relay makes contact to send in an alarm signal. On the other hand, if the monitoring current is interrupted, the relay goes the opposite way and sends in a trouble signal.

A typical rate-of-rise and fixed temperature detector combines the thermostat principle with an air chamber pressing on a diaphragm. This device is recommended where there is apt to be a smoldering fire, one that might develop so slowly that a rate-of-rise detector might not operate, and still has the advantages of this type of detector for a normal fire.

The smoke detector is used where rapid detection is necessary and heat actuated devices are too slow, as in household fire warning systems. These are sometimes called "electronic smoke detectors" and are of two basic types, the photoelectric and the ionization. The photoelectric detector contains a minute electric bulb. If smoke enters, it scatters the light toward a photoelectric cell, triggering the alarm. The ionization type utilizes a small amount of radioactive material to make the air in one or two ionization chambers electrically conductive. This occurs as a result of bombardment of nitrogen and oxygen molecules with alpha particles emitted by the radioactive material. A voltage applied across the ionization chamber causes a small current to flow through the air. This current is more sensitive than light and if smoke (even though invisible) enters the chamber it interrupts the current and triggers the alarm or operates other fire protection devices.

There are many other special type devices on the market and more are being developed, but the operation of a protective signaling system should not depend upon any device that has not been approved by a recognized testing laboratory.

FIGURE 9-5. Ionization type detector on the ceiling protects computer room at Mathematical Applications Group, Inc., Elmsford, New York. Courtesy of the American District Telegraph Company

Spacing of Detectors   All of the standards for automatic detection systems and alarm services call for detectors located on the ceiling (or sidewalls near the ceiling) and installed throughout all parts of the protected premises. This includes all rooms, halls, storage areas, basements, attics, lofts, and all other accessible spaces. Fire records clearly illustrate the danger of attempting to protect only those parts of a building where someone *thinks* a fire may start. It is not unusual to require that detectors be located under large benches or tables; frequently, loading docks and platforms, both underneath and under the cover, should be protected. In short, *detectors should be placed wherever heat may be expected to bank.*

Minimum spacing of detectors depends upon the recommendations of the testing laboratory. Spacing is based upon tests to determine that the detectors, spaced 3.0 m (10 feet) apart, will operate before sprinklers under standard fire conditions. This takes about two minutes and faster operation is desirable at times. To secure this, detectors are installed with closer spacing than is called for by the testing laboratory's report.

The location of smoke detectors should be based upon an engineering survey of the application of this type of protection to the area under consideration. Several things must be taken into consideration: air velocity, travel distance of the smoke, blowers or exhaust fans, air-conditioning systems, temperature variations, and so forth. Care must also be taken to be sure that conditions other than smoke, such as dust, shadows, and the like, will not operate the detectors and signal false alarms.

Other Applications of Fire Detection Principles

From this brief discussion of the principles of detectors, we can readily see that heat actuated devices (sometimes called HADs), using these same principles, can operate any number of fire control devices. We have also seen that the water flow alarm is primarily a local alarm system. But it can be used with any other type of alarm by adding a simple circuit to the place alarms are recorded. Used in this way, a sprinkler system becomes a device for fire detection as well as fire control.

The pneumatic tube principle is used in several forms to operate deluge sprinkler systems, carbon dioxide systems, and dry chemical systems, as well as to operate valves, shut off fan motors, close doors and windows, and other functions. All that is required is the proper mechanical linkage. The thermostat principle is frequently used to control heating devices where a constant temperature is required.

REVIEW QUESTIONS

1. Define a sprinkler system.

2. How effective are sprinklers in extinguishing fires in buildings?

3. Name some of the conditions that require a secondary water supply.

4. How much water is required for a sprinkler system protecting a light hazard occupancy? An ordinary hazard occupancy?

5. What sources of water supply are used for sprinkler systems?

6. What provisions are made for flushing and testing systems?

7. Describe a sprinkler and explain how it works.

8. What is a water flow alarm?

9. How far apart should sprinklers be spaced?

10. Describe the following types of systems:
    (a) Wet pipe system
    (b) Dry pipe system
    (c) Pre-action system
    (d) Recycling system

11. What are outside sprinklers and when are they installed?

12. Where are carbon dioxide systems used? How do they work?

13. Name four types of carbon dioxide systems.

14. Where are foam systems used?

15. Describe a foam-water system.

16. Name three types of expansion foam systems.

17. Describe the combined agent or "twinned" system.

18. What was the first halogenated agent and why was its use discontinued?

19. What two Halons have been adapted to automatic systems?

20. Describe a halogenated system, how it works, and its limitations.

21. Can dry chemical agents be used in automatic systems? If so, how?

22. Describe a fixed water spray system and where it is used.

23. What are the purposes of protective signaling systems?

24. Name the five principles that must be observed for an alarm system to be effective.

25. How are protective signaling systems classified?

26. Describe briefly the following types of systems:
    (a) Local alarm systems
    (b) Remote station systems
    (c) Auxiliary systems
    (d) Central station systems

27. What are the requirements for a reliable household fire warning system?

28. Name three types of fire detectors and give examples of how each works.

10 Municipal Fire Defense

Fire Defense Grading Schedule

In Chapter 3 we saw how the fire protection engineers of the National Board of Fire Underwriters developed a method of inspection and reporting on municipal fire defense systems after the Baltimore and San Francisco conflagrations. After reporting on more than 500 cities throughout the United States, a system for grading cities according to their fire defenses and physical conditions evolved. First published in 1916, this system was called the "Standard Schedule for Grading Cities and Towns of the United States with Reference to Their Fire Defenses and Physical Conditions." The schedule has been brought up to date frequently by revised publication and by supplemental bulletins issued by the National Board and its successor, the American Insurance Association (AIA). When the Insurance Services Office (ISO) took over the Municipal Inspection Service from the AIA in 1970, it also assumed the responsibility for publication of the Schedule. In 1973, ISO published its first revision and changed the name to "Grading Schedule for Municipal Fire Protection."

The Grading Schedule sets standards for various fire defense features. Where the city being graded does not measure up to standard, a certain number of deficiency points is charged. The number of points assessed against a city depends upon the importance of the item being graded and how far the city falls below the standard. The same principle applies to the lack or inadequacy of laws for the control of unsatisfactory conditions, and the way such laws are enforced. Bear in mind that the standards set forth in the schedule are not ideals dreamed up by someone, but are based on actual conditions found in a city in the United States.

How the Point System Works

Table 9 shows the relative values assigned to the various features of a fire defense system. Points indicate the amount of deficiency charged if all of that particular feature is wanting; percent indicates the relative importance of each feature in the entire fire defense system. When these features are subdivided, there are about 100 items to be graded.

TABLE 9. MAJOR CATEGORIES OF A FIRE DEFENSE SYSTEM[a]

FEATURE	PERCENT	POINTS
Water Supply	39	1950
Fire Department	39	1950
Fire Service Communication	9	450
Fire Safety Control	13	650
Total	100	5000

[a]Copyright 1974, Insurance Services Office.

Additional deficiency points may be charged against a system for unusual conditions that may affect fire fighting or contribute to conflagration hazard. These include such items as frequent hot, dry weather, high wind, hurricanes, floods, and earthquakes. Additional deficiencies may be also charged when "water supply" grades considerably better than "fire department," or vice versa. We can easily see that if a fire department is poorly equipped or understaffed, a good water supply is not of as much value as it would be if the department could make the best use of it. Similarly, not even the best fire department can do much without an adequate water supply.

How many deficiency points are assessed is determined, where practical, by applying this simple formula:

$$\frac{\text{Required} - \text{Available}}{\text{Required}} = \text{Percent Deficiency},$$

where Required is the number or amount set by the standard and Available is what actually exists.

For example: according to the Schedule, a certain city should have four pumpers to meet the standard, but it has only three. Using the formula yields the following results:

$$\frac{4-3}{4} = 0.25, \text{ or } 25\%.$$

Grading engineers convert this percentage into the number of deficiency points assessed by using the Deficiency Scale that appears in Table 10.

To use the scale, first read from left to right along the top line, "Deficiency Percentage in Tens." Using the example, stop at the 20% column and then read down the Deficiency Percentage in Units axis to 5%. The resulting figure, 35, is the number of deficiency points that will be assessed for the substandard number of pumpers in our example.

Notice, though, that the number of deficiency *points* to be charged is not the same as the deficiency *percentage* but considerably higher. Notice also that the number of deficiency points assessed does not increase in direct proportion to the deficiency percentage; instead, there is a relatively higher charge imposed as deficiency percentage increases. Look at Table 10 once again, we can readily see that a deficiency of 10 percent draws only 10 deficiency points but a 25 percent deficiency draws a penalty of 35 deficiency points. Similarly, a 50 percent deficiency draws a 90 point penalty and a 100 percent deficiency costs 200 deficiency points.

TABLE 10. DEFICIENCY SCALE[a]

		0%	10%	20%	30%	40%	50%	60%	70%	80%	90%	100%
					DEFICIENCY PERCENTAGE IN TENS							
DEFICIENCY PERCENTAGE IN UNITS	0%	0	10	25	45	67	90	112	134	156	178	200
	1%	1	12	27	47	70	92	114	136	158	180	
	2%	2	13	29	50	72	94	116	138	160	182	
	3%	3	15	31	52	74	97	119	141	163	185	
	4%	4	16	33	54	77	99	121	143	165	187	
	5%	5	18	35	57	79	101	123	145	167	189	
	6%	6	19	37	59	81	103	125	147	169	191	
	7%	7	21	39	61	83	105	127	149	171	194	
	8%	8	22	41	63	85	108	130	152	174	196	
	9%	9	24	43	65	88	110	132	154	176	198	

[a]Copyright, 1974, Insurance Services Office.

There is a reason for this extra penalty: the increased number of deficiency points charged reflects the difference in effect on a given fire defense system of a slight or a serious deficiency.

The Schedule also provides for the use of fractions or multiples of the Deficiency Scale, according to the importance of the item being rated. For example, if a department with six companies had two officers less than were needed to have an officer on duty at all times with each company, the grading engineer would first apply the formula:

$$\frac{18 \text{ (Required) minus 17 (Available)}}{18 \text{ (Required)}};$$

this fixes the percentage of deficiency at 11 percent. Using the Deficiency Scale to convert, the grading engineer then arrives at a charge of 12 deficiency points but the Schedule sets the maximum charge for this deficiency at one-fourth that amount, or 3 deficiency points.

The following example illustrates the application of a multiple of the deficiency scale: Grading engineers determine the required fire flow for a given city as 13 250 liters per minute (3500 gallons per minute) for 10 hours, or a total of 8.0 megaliters (2.1 million gallons). City records show average daily consumption is 19.1 megaliters (5.04 million gallons), or 8.0 megaliters (2.1 million gallons) in 10 hours. For the system to be rated as completely adequate, it must be able to supply both these demands simultaneously, a total of 15.9 megaliters (4.2 million gallons). Inspection and testing discover that the system will provide only 11 megaliters (3 million gallons) during a 10-hour period, or 4.5 megaliters (1.2 million gallons) less than required.

$$\text{Using the formula, } \frac{4.2-3}{4.2} = 0.285, \text{ or 29 percent, and}$$

using the Deficiency Scale, this percentage draws an assessment of 43 deficiency points. However, for this deficiency, the Standard Schedule, under Water Supply, Item 1, requires an assessment one and a half times the deficiency scale rating, or 44 deficiency points.

Since it is not always practical to determine a percentage, some items are charged a fixed number of deficiency points under certain conditions. This fixed number, however, only serves as a guide for the engineer who is applying the schedule. He or she may assign a greater or smaller number of points, depending on the seriousness of the condition he finds. A guide is also provided for establishing percentage of deficiency according to "engineer's judgment." This is based on whether conditions are good, fair, or poor or whether the deficiency is slight, moderate, considerable, or serious.

When all items have been graded, deficiency points are totaled. This total determines *the relative protection class* of the municipality in question. If deficiency points total between 0 and 500, the city is First Class.

So far, no city graded enjoys this distinction. A city with a total between 501 and 1000 points is Second Class; 1001 to 1500 is Third Class. Classification goes up one number for each 500 deficiency points assessed. A city grading more than 4500 points is Tenth Class, or unprotected. There are some exceptions to this rule in the Ninth and Tenth classes: For example, a Ninth Class city is normally one that grades between 4001 and 4500 points, but a city grading less than 4001, but without a recognized water system, is also classed as nine. A city grading Tenth Class with a fire department and no recognized water system, or with a water system but no recognized fire department, is a Tenth Class city.

A statement regarding these classes, which is to be incorporated into the introduction of the forthcoming edition of the Municipal Grading Schedule, is interesting and pertinent to our study of the Schedule:

> The insurance classification developed under the schedule is only one of several elements used in the development of fire insurance rates. Although the schedule provisions may be of assistance to municipal officials when used in conjunction with their analysis of local needs, capabilities and priorities, the schedule is not intended to serve as a primary planning guide for local fire protection. Recommendations in connection with insurance classifications should be helpful to municipal officials when reviewed in connection with more specific studies of local needs by consultants, staff, or local task forces in arriving at fire protection decisions based upon analysis of local priorities and financial capabilities.*

Individual features of a fire defense system may also be classified, just like the entire municipality. This is done by dividing the deficiency points charged against a given feature by 10 percent of the points allocated to it. If the result is a fraction or decimal, the next higher whole number is the class. For example, a fire department rated at 620 deficiency points would be designated as Fourth Class:

$$\frac{620-\text{deficiency charged}}{195-\text{ten percent of the 1950}} = 3.2$$
$$\text{points allotted to Fire}$$
$$\text{Department (See Table 9)}$$

Since this is a decimal, the next *highest* figure, Class 4, is the correct rating.

*Quoted from a paper presented at the 46th annual Fire Department Instructors Conference at Memphis, Tennessee, March 1974, by Weldon F. Williams, Executive Vice President of Insurance Services Office.

Fire Defense System Features

Water Supply System

Since a water supply is considered one of the most important features of a fire defense system, let us consider how a water supply system operates. Water systems can be publicly or privately owned. When they are public, they are managed by a department of municipal government, or by an agency such as a utility district serving several municipalities. Regardless of ownership, successful operation requires a well qualified superintendent or chief engineer and assistants. They should also have the proper professional education. In general, having competent people with long service results in an efficient organization.

Complete records of the system are essential. These records should include complete plans of the supply works, maps of the distribution system showing location of mains, valves, hydrants, and so forth, and records of consumption, storage levels, gate valve and hydrant inspections, and all other operations of the system. Detailed maps of the distribution system should be available for maintenance crews for accurate information in case of emergency. All records must be kept up to date in convenient form, suitably indexed, and safely filed.

Emergency crews, with radio-equipped transportation and the tools to make emergency repairs to the system, should be available on short notice.

A close working relationship between the fire department and the water supply organization is required if the fire department is to make the best use of the distribution system. Representatives of the two departments should consult on hydrant locations and the amount of water needed for fire protection. The water department should also receive all alarms of fire so that it can put additional equipment in service, raise pressures, or operate emergency valves. In case of second alarms or serious fires, someone from the water department should respond, prepared and equipped to operate valves, make emergency repairs, or to assist the fire department to use the water system most effectively.

Adequacy   A water system is considered adequate if it can deliver the required fire flow for a specified number of hours in addition to the amount of water used for all other purposes during the period of peak demand (maximum daily consumption rate). Consumption rate is· determined by the largest amount used during any 24-hour period during the past 3 years. Special use, such as refilling a reservoir, will not be counted. *Fire flow is the amount of water available for fire fighting purposes, not necessarily from one hydrant, but from several hydrants in a particular locality or district.* Required fire flow, or the amount of water that might be needed to extinguish a serious fire or conflagration in a particular district, involves so many factors that there is no simple formula to determine it. Before the days of supermarkets, shopping centers, and the decentralization of cities, a formula for fire flow requirements in mercantile districts of cities, based upon population, was developed by National Board of Fire Underwriters engineers.

Using their method, figures indicated that an average city with a population of 1000, would require 3785 liters (1000 gallons) of water *per minute* to protect its mercantile district; a city of 4000 would require 7570 lpm (2000 gpm); 10,000 population, 11 360 lpm (3000 gpm); and so on, up to 200,000 population which would require 45 420 lpm (12,000 gpm). This formula worked fairly well until large area supermarkets began to appear in all of our cities. Fires in these structures showed that estimates of required fire flow were inaedquate, particularly when applied to smaller communities.

Judgment, based upon experience and circumstances, is used to determine the needed fire flow. Experienced fire department officers sometimes include an estimate as part of preplanning. What they do is to visualize a serious fire at the time the fire department arrives. Knowing the number of hose lines and powerful streams they would need, they can make a good estimate of the amount of water that would be required. Other methods of estimating fire flow requirements consider type of construction, volume and contents of buildings in the area, and exposures that will have to be protected. There are formulas that incorporate these factors, but all of them require judgment in their application.

In 1974, the Insurance Services Office published a "Guide for Determination of Required Fire Flow" in which a formula was given using factors for ground area and type of construction. From this, tables have been prepared showing required fire flow for buildings of various areas and heights up to six stories for three different types of construction: wood frame, ordinary, and fire resistive. This is just the beginning, because the fire flow figure obtained from the table must be increased or decreased for the relative hazard of the occupancy, decreased if the building is equipped with automatic sprinklers, and increased for exposures. The foreword to the guide states, "It should be recognized that this publication is a 'guide,' in the true sense of the word, and requires knowledge and experience in fire protection engineering for its effective application." With that in mind, it is a very helpful publication.

As a rule of thumb, however, we can use the following as a guide to required fire flow:*

*The following includes copyrighted material of the Insurance Services Office, with its permission. Copyright 1974, Insurance Services Office.

- Well-spaced, average, single-family dwellings: 1.89–3.785 klpm (500–1000 gpm)

- Two- to three-story apartment houses, well-spaced: 5.678–9.462 klpm (1500–2500 gpm)

- Neighborhood shopping centers, ordinary construction, with a 20,000 square-foot supermarket: 11.36–15.14 klpm (3000–4000 gpm)

- Large regional shopping center, covering many acres: 17.03–22.71 klpm (4500–6000 gpm)

- Industrial park district, one-story buildings of moderate area and buildings usually sprinklered: 11.36–15.14 klpm (3000–4000 gpm)

The minimum fire flow requirement is 1.89 klpm (500 gpm) and the maximum for a single fire is 45.42 klpm (12,000 gpm), but if a system cannot deliver at least 8.946 klpm (250 gpm) for two hours it cannot be recognized as a fire defense system.

Whenever we estimate fire flow, we make allowance for water wasted at the fire by broken services, ineffective hose lines, and so forth.

The length of time fire flow should be sustained also depends upon the same conditions that determine the quantity of water required. Table 11 shows the standard, expressed in terms of fire flow required.

Available fire flow can be determined by opening one or more hydrants in a district, measuring the flow from each, and at the same time measuring the drop in pressure in the mains. We can then determine, either by computation or from tables, the amount of water available at any pressure.

TABLE 11. REQUIRED DURATION FOR FIRE FLOW[a]

REQUIRED FIRE FLOW		REQUIRED DURATION
klpm	gpm	hours
9.5 and less	2500 and less	2
9.5–13.3	2500–3500	3
15.2–17.1	4000–4500	4
19.0–20.9	5000–5500	5
22.8–24.7	6000–6500	6
26.6–28.5	7000–7500	7
30.4–32.3	8000–8500	8
34.2–36.1	9000–9500	9
38.0 and more	10,000 and more	10

[a]Copyright 1974, Insurance Services Office.

To overcome friction loss in the hydrant branch, the hydrant, and the suction hose to the pumpers, fire flows generally require a residual pressure in the mains of 138 kPa (20.0 psi). Where pumpers are not available or when for other reasons the fire department depends upon direct hydrant streams for water supplies, residual pressures of 517 kPa (75.0 psi) are needed.

When we study a water system from the fire defense standpoint, we determine what features of the system may cause restriction of the fire flow in a particular district. Such features include:

- inadequate source of supply, such as limited production from wells or inadequate storage of stream waters in reservoirs;
- a defect in the supply works, such as inadequate filter capacity, insufficient pumping capacity, or supply lines too small to deliver the required amount of water to the district;
- faulty design of the distribution system.

We will consider some standards for water systems a little later in this chapter.

To determine how serious a deficiency is, we must compare required fire flow with the flow available at the time of both average and maximum consumption. Obviously, a system that can deliver the required amount of water for a fire during average conditions, but is limited at times of heavy consumption demands, is better than one that is inadequate at all times. But no system can be considered adequate unless it can meet fire demands under all conditions.

One other element to consider in measuring the adequacy of a water system's ability to provide fire protection is hydrant distribution. How many hydrants are needed depends upon required fire flow and the area to be served, as shown in Table 12.

Table 12 needs some explanation, however. Primarily, it establishes the average number of hydrants required based upon the fire flow. Indirectly, it controls the spacing of these hydrants but not their location. For example, consider a high-value district measuring six blocks on a side with a required fire flow of 45.42 klpm (12,000 gpm). To meet the Standard, this district would have to have 36 hydrants. We reach this figure as follows: an average city block measures 61.0 × 61.0 m (200 × 200 feet) and contains 3.72 km² (40,000 square feet). There are 36 such blocks in a six block square; hence, 36 hydrants are needed to protect the area. At the same time, Table 12 limits the area any one hydrant can cover to 3.72 km² (40,000 square feet). As a result, there must be at least one hydrant in each block of the district in our example to meet the Standard.

In practice, hydrants are spaced a good deal closer than one to a block, particularly in high-value districts. The type of hydrant most commonly in use today, with two 2½-inch outlets and one 4- to 6-inch pumper outlet at pressures meeting Standard Schedule requirements, will only deliver 2.27 klpm (600 gpm). Consequently, to make full use of modern apparatus, and to keep hose lines as short as possible, cities commonly locate hydrants in high-value or congested districts about 61.0 meters (200 feet) apart, or, roughly, one on each corner of a block. In residential districts, depending on construction density and structural conditions, hydrants are spaced further apart, location often conforming exactly to the area requirements established by Table 12.

TABLE 12. STANDARD HYDRANT DISTRIBUTION[a]

FIRE FLOW REQUIRED		AVERAGE AREA PER HYDRANT	
lpm	gpm	square feet	square meters
3 785	1,000 or less	160,000.	14 864
5 678	1,500	150,000.	13 935
7 570	2,000	140,000.	13 006
9 462	2,500	130,000.	12 077
11 360	3,000	120,000.	11 148
13 250	3,500	110,000.	10 219
15 140	4,000	100,000.	9 290
17 030	4,500	95,000.	8 826
18 930	5,000	90,000.	8 361
20 820	5,500	85,000.	7 897
22 710	6,000	80,000.	7 432
24 600	6,500	75,000.	6 968
26 500	7,000	70,000.	6 503
28 390	7,500	65,000.	6 039
30 280	8,000	60,000.	5 574
32 170	8,500	57,500.	5 342
34 065	9,000	55,000.	5 110
37 850	10,000	50,000.	4 645
41 640	11,000	45,000.	4 181
45 420	12,000	40,000.	3 716

NOTE 1:   If hydrants are so spaced that the areas served in some districts or in portions of some districts are considerably larger than the average, an additional percentage deficiency shall be assigned. The fact that the average area served meets the requirement shall not preclude the assignment of such additional deficiency.

NOTE 2:   Where it is the practice of the fire department to connect two pumpers to a hydrant, each hydrant with two pumper outlets may be counted as 1½ hydrants. Capability of hydrant, street connection, and street main to supply two pumpers shall be considered.

[a]Includes copyrighted material of Insurance Services Office with its permission. Copyright 1974, Insurance Services Office.

Reliability   In designing a water system to be used for fire defense, we must do more than provide for adequate water *under normal conditions.* Studies of conflagrations indicate that very often something went wrong with a water supply just when it was needed most. Frequently, this failure is one reason the fire spread. Since water is the fire fighter's chief weapon, there should be some way to compensate for interruptions of water supply during an emergency. Whether a water system has such a compensating feature or not determines its reliability.

Starting at the source of supply, some of the things that can cause a break in service are drought, forest fire, earthquake, ice formation, and silting or shifting of channels. Where the supply comes from wells, we face the possibility of sanding, intrusion of salt water, heavy drawdown from too many wells in the area, or even the interruption of underground supply by earthquakes. Of course we can do little to stop these forces of nature, but we can compensate for them by providing storage capacity to meet demand during critical periods or we can arrange for emergency supplies to be available.

The most reliable water supply is gravity-fed from storage at high elevation. Since this is not available to most communities, we depend upon pumps to force the water to elevated storage and provide the needed pressure. To meet reliability standards under these conditions, pumping capacity must be great enough to provide the required fire flow for the specified time over a period of five days, with consumption at the maximum daily consumption rate, and with the two *largest capacity pumps* out of service. Elevated storage, if available, can be taken into account in determining the ability of the pumping system to meet this standard of reliability. Some systems depend upon two or more stages of pumping, such as deep-well pumps, delivering water to a surface reservoir, and high-lift pumps, taking suction from the surface reservoir and pumping into the system. In such cases, each stage of pumping should provide the duplication outlined above.

A common cause of disruption of water supply during a fire is electric power line failure. To counteract this threat, the Standard states, "Electric power supply shall be so arranged that a failure in any power line or the repair or replacement of a transformer, high-tension switch, control unit, or other power device will not prevent the delivery, in connection with storage, of the required fire flow for the specified duration during a period of two days with consumption at the maximum daily rate." This indicates complete duplication of power lines, but this is not enough to meet strict reliability standards. Power lines should run underground from the substation of the utility company to the pumping station to avoid exposure to all the hazards of overhead transmission lines.

When the prime movers for the pumps are steam or internal combustion engines, similar precautions must be taken in the interest of reliability. Steam boilers, for instance, working at one-quarter of the entire capacity and with one boiler out of service, must be sufficient to operate all machinery and pumps needed to deliver the required quantity of water. Steam must be kept at least to one-half the required pressure for the pumps to deliver the necessary fire flow and consumption for a period of two hours. Enough fuel for at least a five-day supply for boilers or internal combustion engines should be kept safely stored underground. Special attention must be given to the effect of repairs or replacement of any fuel pump, any valve, or failure in any line. Any of these must not prevent the delivery of required fire flow and consumption for two days.

To be sure that all the conditions, arrangements, and operations of the pumping plant equipment meet reliability standards, we must take a pessimistic approach. We must try to anticipate everything, including the operating personnel and their ability to make repairs, that might interrupt the supply, and then guard against it. We must also consider general arrangements, spare parts available, the possibility of the plant flooding, purification works, and arrangements for by-passing filters, if necessary. The list could go on and on, but we cannot stop with plant equipment.

There is always the possibility of losing one or more pumping plants because of fire. Ideally, all plants would be of fire-resistive construction, contain no combustible material, and be equipped with automatic sprinklers. Frequently, pumping plants contain garages, repair shops, offices, and so on. In cases of this kind, the pumping equipment should be cut off from the rest of the building by a fire-resistive wall or partition. First-aid fire protection such as hand extinguishers and standpipe lines with small hose should be provided at every pumping plant. Fire hazards should be reduced to a minimum. Where possible, we should locate pumping plants where they are not exposed by other buildings; when this cannot be done, suitable protection should be provided by outside sprinklers, fire doors, wired glass windows, or other approved methods.

Next, from the reliability standpoint, are the mains. These include all piping, fittings, conduits, and mains that supply water to the system from the source of supply. It applies to both gravity and pumping systems. We start by using a map of the system to fix the location of the most serious break that could occur in any main. We determine seriousness by comparing the amount of water normally available or required for fire flow to what would be available if a particular break should occur. The mains are considered reliable if the remaining pipes can supply required fire flow and maximum consumption for the required length of time. If they cannot, the number of places or the length of main in which such a serious break could occur must be considered when judging the reliability of mains.

Types of material used and method of installation are also important factors in determining the reliability of both supply mains and all mains in the distribution system. The best type of material for pipelines depends on many factors, including size of the conduit, soil conditions, terrain, and installation facilities. Installation itself is extremely important. Special construction is called for at stream crossings, railroad crossings, bridges, or in filled ground, with care taken that mains do not endanger each other in case of a break or failure of any kind. All mains, except supply mains on private right-of-ways, should be at least two feet below the surface of the ground to protect them from traffic injury; greater cover may be required to prevent freezing. A review of service records will show the frequency of leaks, breaks, and repairs, giving an indication of the reliability of the installation of mains.

Design   Some features of the design of a water system give us an indication of the efficiency of the distribution system for fire protection, since they generally affect both adequacy and reliability. The arterial system that connects with the supply mains and secondary feeders should extend throughout the distribution system and be large enough to meet the fire flow and consumption demands of any district. No large section of a community should be dependent upon a single main for supply. To avoid this, the arteries (or secondary feeders in large systems) should be spaced no more than 915 meters (3000 feet) apart and should be looped, tied together at intervals with large mains, so water may flow in either direction.

Minor distribution mains should form a gridiron of pipe, not smaller than a 15-cm (6-inch) inside diameter, and so arranged that the longest side of the gridiron does not exceed 183 meters (600 feet). Dead-end pipelines supplying hydrants are poor practice. Where a dead end cannot be avoided or where a long gridiron is necessary, the system should use 20-cm (8.0-inch) or larger pipe.

Design and development of a water system should consider future developments that may require large quantities of water, either for daily use or fire protection, such as industrial areas or large shopping centers with very large areas. Fire flow requirements in these districts are often greater than those for the principal business district. Another design consideration is having adequate water at sufficient pressure for automatic sprinkler systems in any high-value district. This supply should be available even at times of heavy draft upon the system.

In a well-designed water system, fire hydrants will be low in friction loss, and in compliance with American Water Works Standard for Dry Barrel Hydrants (AWWA-C 502). Well-designed wet barrel hydrants are satisfactory in mild climates. They should have at least two 2½-inch outlets, as well as a large outlet for a pumper connection. This requires a connection to the street main at least 15 cm (6 inches) in diameter, equipped with a gate valve to completely shut the hydrant off from the water system.

Fire Department

The most adequate and reliable water system in the world is of little value for fire protection without an organization equipped to get the water onto the fire. Equipment involves so much of importance to fire science that we devoted all of Chapter 5 to this subject. Here, we will briefly consider the general standards affecting fire departments as part of a municipal fire defense system.

Organization   For a fire department to be "recognized," it must meet five conditions:

- It must be organized on a permanent basis, under applicable state and local laws, and must include some person, designated by election or appointment, who is responsible for operations at fires.

- It must have enough active members to provide at least four persons in answer to any alarm.

- An appropriate training program must be maintained for all active members.

- Response to alarms must include at least one piece of apparatus, suitably designed and equipped for fire service; proper maintenance and housing must be provided for the apparatus.

- Reliable means must be provided for the receipt of alarms of fire and transmission of the alarm to fire fighters.

Notice that these are *minimum requirements for recognition.* The lack of any of these five items justifies a refusal to recognize the organization as a fire department.

The chief of a fire department should be qualified by experience and trained in administration. The chief should be appointed for an indefinite term and be removed only for cause, after a proper hearing. If there are more than two fire companies in a department, there should be an assistant chief who takes charge in the absence of the chief. Battalion or district chiefs should be appointed for departments with more than eight companies. There should be enough company officers for one to be on duty *at all times* with each company. Company officers should also be competent and experienced in the fire service. There should .be an operator and driver on duty at all times for each piece of apparatus in service.

Company Strength   To obtain normal efficiency from the apparatus, six fire department members, including company officers, are required to be on duty with each required engine and ladder company. That is called the "required strength"; the "on duty strength" is the number of members on duty during vacation periods less the average number on detail and sick leave.

Consideration is given to departments having fire fighters paid on call or volunteers, provided they have adequate means of receiving alarms, such as public alarms, radios, or fire alarm tappers in their homes or places of business. To compute the available personnel for such departments, we determine the average number of fire fighters responding to fires. This average, divided by four, gives us the personnel equivalent of full-time paid fire fighters, but in a recognized department, the total of these equivalents cannot exceed one-third of required company strength. The standard also takes into consideration off-duty fire fighters and companies from other departments, responding as outside aid to serious fires, as a partial offset to deficiencies in personnel on duty.

As with all emergency organizations, a fire department must have discipline and training to be efficient. These subjects were completely discussed in Chapter 4. At this point, we only need to point out that *printed regulations* for the control of the department should be in the hands of every member; the chief should have authority to enforce these regulations, subject to review by the supervising body. A department should have training facilities and a training program for all company members, in charge of a competent officer, supplemented by daily drills at the fire stations.

Number and Distribution of Fire Companies   Like requirements for fire flow, the number of companies needed for adequate protection in a fire department were based upon the population of the city or fire protection district. That method is no longer used in the Grading Schedule since the physical conditions of cities have changed because of decentralization and, instead, the number and location of fire companies are based upon the fire flow requirements in any part of the city.

Table 13, taken from the Grading Schedule, indicates the number of companies needed within the maximum travel distance based upon required fire flow. As an example of how the table is used, assume a shopping center where we have determined the required fire flow to be 30.28 klpm (8000 gpm). A glance at the table shows there should be one engine company within 1.6 km (1.0 mile) and one ladder company within 2.4 km (1½ miles). To provide adequate first alarm response there must be an additional engine company within 2.4 km (1½ miles); since only one ladder company is required on first alarms, the first company due to respond is all that is needed. For multiple alarms, however, there should be seven more engine companies (to make a total of nine) within 7.2 km (4½ miles) and two more ladder companies (to make a total of three) within 5.6 km (3½ miles).

TABLE 13. NUMBER OF ENGINE AND LADDER COMPANIES NEEDED WITHIN TRAVEL DISTANCE OF ESTABLISHED FIRE FLOW[a]

FIRE FLOW		FIRST CO. DUE						FIRST ALARM						MAXIMUM MULTIPLE ALARM					
		Engine			Ladder			Engine			Ladder			Engine			Ladder		
klpm	gpm	No.	Km	Mi.	No.	Km	Mi.	No.	Km	Mi.	No.	Km	Mi.	No.	Km	Mi.	No.	Km	Mi.
less than 7.57	2000	1	2.4	1.5	1[b]	3.2	2.0	2	6.4	4.0	1[b]	3.2	2.0	2	6.4	4.0	1[b]	3.2	2.0
7.57	2000	1	2.4	1.5	1[b]	3.2	2.0	2	4.0	2.5	1[b]	3.2	2.0	2	4.0	2.5	1[b]	3.2	2.0
9.46	2500	1	2.4	1.5	1[b]	3.2	2.0	2	4.0	2.5	1[b]	3.2	2.0	2	4.0	2.5	1[b]	3.2	2.0
11.36	3000	1	2.4	1.5	1[b]	3.2	2.0	2	4.0	2.5	1[b]	3.2	2.0	3	4.8	3.0	1[b]	3.2	2.0
13.25	3500	1	2.4	1.5	1[b]	3.2	2.0	2	4.0	2.5	1[b]	3.2	2.0	3	4.8	3.0	1	3.2	2.0
15.14	4000	1	2.4	1.5	1	3.2	2.0	2	4.0	2.5	1	3.2	2.0	4	5.6	3.5	1	3.2	2.0
17.03	4500	1	1.6	1.0	1	3.2	2.0	2	3.2	2.0	1	2.4	1.5	4	5.6	3.5	2	4.0	2.5
18.93	5000	1	1.6	1.0	1	3.2	2.0	2	3.2	2.0	1	2.4	1.5	5	5.6	3.5	2	4.0	2.5
20.82	5500	1	1.6	1.0	1	2.4	1.5	2	3.2	2.0	1	2.4	1.5	5	5.6	3.5	2	4.0	2.5
22.71	6000	1	1.6	1.0	1	2.4	1.5	2	3.2	2.0	1	2.4	1.5	6	6.4	4.0	2	4.0	2.5
24.60	6500	1	1.6	1.0	1	2.4	1.5	2	2.4	1.5	1	2.4	1.5	6	6.4	4.0	3	5.6	3.5
26.50	7000	1	1.6	1.0	1	2.4	1.5	2	2.4	1.5	1	2.4	1.5	7	6.4	4.0	3	5.6	3.5
28.39	7500	1	1.6	1.0	1	2.4	1.5	2	2.4	1.5	1	2.4	1.5	8	7.2	4.5	3	5.6	3.5
30.28	8000	1	1.6	1.0	1	2.4	1.5	2	2.4	1.5	1	2.4	1.5	9	7.2	4.5	3	5.6	3.5
32.17	8500	1	1.6	1.0	1	2.4	1.5	2	2.4	1.5	1	2.4	1.5	9	7.2	4.5	4	6.4	4.0
34.07	9000	1	1.2	.75	1	1.6	1.0	3	2.4	1.5	2	3.2	2.0	10	7.2	4.5	4	6.4	4.0
37.85	10000	1	1.2	.75	1	1.6	1.0	3	2.4	1.5	2	3.2	2.0	12	8.0	5.0	5	7.2	4.5
41.64	11000	1	1.2	.75	1	1.6	1.0	3	2.4	1.5	2	3.2	2.0	14	8.0	5.0	6	8.0	5.0
45.42	12000	1	1.2	.75	1	1.6	1.0	3	2.4	1.5	2	3.2	2.0	15	8.0	5.0	7	8.0	5.0

[a]Copyright 1974, Insurance Services Office.
[b]Where there are less than 5 buildings 3 stories in height, a ladder company may not be needed to provide ladder service.

By applying these standards to various parts of the city, the total number of engine and ladder companies can be determined and that total determines the size of fire fighting force.

Considering the complexity of determining the number and distribution of companies needed for any community, we can readily see that it can only be done after a careful study of all local conditions by someone familiar with the Standard.

Equipment   The National Fire Protection Association's Standard No. 1901, *Automotive Fire Apparatus,* is used to determine the equipment required and the suitability of pumpers, ladder trucks, and other apparatus. The capacity of the pumpers responding to multiple alarms in any district should be at least equal to the required fire flow in that district. Where a district has a high pressure water system for fire protection, the amount of water available at 1030 kPa (150 psi) residual pressure may be counted as pumper capacity and some engine companies may be replaced with hose companies. Each engine or hose company should be equipped with 731.5 m (2400 feet) of 2½-inch or larger hose, 183 m (600 ft.) of 1½-inch hose, and 61.6 m (200 ft.) of booster hose. There should be a reserve pumper, fully equipped, for every eight pumpers required.

Each ladder company should be provided with a ladder truck, preferably of the aerial or elevating platform type. Tools and equipment should include those listed in NFPA Standard No. 1901. In districts having no buildings over three stories in height and not required to have a standard ladder truck, the engine companies or special apparatus should carry sufficient ladders to reach the roofs of all buildings. There should be a reserve ladder truck for every five ladder trucks required.

A pumper ladder truck (a truck equipped with a complement of ground ladders and other ladder truck equipment and with a pump and hose) is counted as a pumper and one-half a ladder company, or as a ladder company and one-half an engine company, depending upon where it is most needed.

Special Equipment   Powerful stream appliances should be provided in municipalities having buildings of large area or structures with three stories or more. These appliances should include such items as turret or monitor nozzles on apparatus, deluge sets or portable monitors and large spray nozzles. For possible fires in some districts, cellar pipes, distributing nozzles, foam and other equipment are needed. A complete ladder pipe assembly, either permanently installed or readily attached, should be provided with each aerial ladder truck.

In cities having a waterfront with an occupied wharf frontage of at least one mile, a fire boat should be part of the fire department apparatus. No such frontage should be more than 2.4 km (1½ miles) from a fire boat. The pumping capacity of an individual fire boat should not be less than one-half the required fire flow, and preferably not less than 18.93 klpm (5000 gpm). Other conditions that may require special equipment in a municipality include extensive brush areas, excessive bulk oil and other hazardous storage, and so forth.

Although not specifically mentioned in the Schedule, NFPA Standard No. 1901 calls for salvage covers on pumper and ladder trucks. In large cities special salvage companies are provided near high-value districts.

Maintenance is an important factor with all fire department apparatus and equipment. The engineers who make the inspections for grading pay close attention to the condition of the hose, for instance, which should be tested annually to at least 1720 kPa (250 psi). Suitable facilities should be provided for washing and drying hose after use. Pumpers are tested to be sure they can deliver rated capacity at standard pressures.

Operations   All phases of operation at fires are considered, including the ability of the department to operate effectively at both small and large fires using modern fire methods. The inspecting engineers observe how rescue work is performed, the ability to maintain radio communications with all companies on the fire grounds, proper use of private fire protection systems, and all other factors that adversely or favorably affect fire operations.

To have good fire operations requires training of all personnel and that, in turn, requires facilities of adequate size and suitably equipped for proper instruction. This calls for a drill tower, classrooms, and instruction aids. The training program should be under the supervision of a competent training officer and should include the study and development of modern practices and procedures as well as practice in ladder and hose drills. The training program should also include frequent building inspections by each company for the purpose of prefire planning. A fringe benefit of inspections is the familiarity with buildings in the district. This can be of great help when fire fighters are working in the dark or in smoke.

Conditions that adversely affect operations should also be taken into consideration. These include such factors as incompetent supervisory personnel; unsatisfactory appointment, promotion, and retirement provisions; frequent changes in supervisory personnel; inadequate rules and regulations, authority of the chief, and discipline; possibility of strikes; poor location of and hazards in fire stations; inadequate provision for refueling apparatus; and inadequate records. Physical conditions that might cause delay in response to alarms should be considered in this category. These may include poorly paved or narrow streets, steep grades, poor parking regulations, and many other things. Other factors may interfere with fire fighting operations; for example, overhead wiring in streets or alleys can obstruct handling ladders and hose streams, and high-voltage lines can be dangerous.

Fire Service Communications   As in any operation that requires teamwork, communications are essential to the fire defenses of any municipality. In a fire department this element is important in two ways: (1) to notify the fire department that there is a fire and where it is, and (2) to keep all companies in contact with the chief officer and each other. To provide the first we must have a municipal fire alarm system whose function is to receive alarms from the public and transmit them by signal to the fire companies. To provide for the second function radio and telephone is provided. The Grading Schedule considers both these under the heading of Fire Department Communications.

Fire Alarm Systems, General   The standards used in the Schedule can be found in NFPA Standard No. 73, *Public Fire Service Communications.* This standard was first published by the National Fire Protection Association in 1898, so the need for a reliable fire alarm system is not exactly new. In small communities the alarm system often consists of the number of a telephone where someone is available at all times and can sound a public alarm such as a siren or an air horn. This calls out the volunteer fire fighters and usually a cloud of smoke indicates the location of the fire. The standard deals with more complex systems. We shall not cover all of the details, however, but will touch on the items covered by the schedule.

Fire alarm telegraph systems are the most widely used systems and the oldest. These are divided into two types: Type A systems are those in which an alarm from a box is received and then retransmitted to fire stations either manually or automatically. This type of system can be used in any municipality, but is required when the emergency calls from boxes exceed 2500 per year. Type B systems are those in which an alarm from the box is automatically transmitted directly to the fire station.

The series telephone system is one in which the box contains a telephone instead of a coding mechanism and several of the telephones are connected in series (i.e., on the same circuit).

The parallel telephone system has telephones in the boxes, but has an individual circuit to each box.

In the coded radio system, the boxes contain a radio transmitter which sends a coded signal to a communication center, much the same as a fire alarm telegraph box, but by radio instead of by wire circuits.

Communication Center   This is the building or portion of a building that houses the equipment upon which the receipt and transmission of alarms are dependent. Under ideal conditions the communication center is in a fire resistive building unexposed within 45.7 m (150 feet). If this is not possible and if the equipment is housed in a building with other occupancies, safeguards should be taken to cut off the center from all outside hazards and to avoid placing any combustible material in the structure or furnishings. Consideration should also be given to protection against disruption of service because of earthquakes, floods, or other catastrophes.

In large systems it is sometimes desirable to terminate some circuits in satellite communication centers for the purpose of operational efficiency. These centers should have the same safeguards as the central communication center, with one operator on duty at all times.

Equipment in communication centers must be adequate and reliable so as to insure transmission and recording of all alarms, even those by telephone. The type of equipment depends upon the type of alarm system; these are described in NFPA Standard No. 73. To assure reliability all wired circuits, including the power supply upon which the transmission and receipt of alarms depends, must be under constant electrical supervision. That is, they must be so designed as to give a prompt warning of conditions adversely affecting reliability. Trouble signals must actuate a sounding device in an area where someone is always on duty and must be distinct from alarm signals.

The number of operators required to be on duty at communication centers depends upon the number of alarms and trouble signals received. Operators should be competent, familiar with the system, and in good health.

Equipment in and Circuits to Fire Stations   Normally two separate dispatch circuits are required for transmitting alarms to fire stations, one of which must consist of (1) a supervised wire circuit, (2) a radio circuit with duplicate base transmitter, or (3) a supervised carrier circuit (microwave). In type B systems where there are less than 600 alarms per year and all fire stations have recording and sounding devices responsive to each box circuit, the second dispatch circuit is not required and when a radio circuit is used as a dispatch circuit, the transmission of an alarm must be preceded by a distinctive tone.

Each fire station must have two separate and distinct facilities for receiving notification from the communication center that response of apparatus and crews is expected and means must be provided in the fire stations for acknowledging receipt of the alarm. This can be done by radio or telephone. The type of audible alarm devices necessary depends upon whether watch is maintained at all times in the station, but arrangements must be such that fire fighters can be alerted at any time of the day or night.

Boxes   Fire alarm boxes available to the public should be of a distinctive color, readily recognizable, and operable in all kinds of weather. When a wire is broken, for instance, the system should still be able to transmit the signal. Fire alarm telegraph and coded radio boxes should be designed to transmit at least three rounds of the box number, and radio boxes should transmit one round of "test" or "tamper" signals. When two or up to four boxes are operated at the same time, the system should transmit one box number alarm after another. They should also give an audible or visual indication to the user that the box is operating or that the signal has been received.

In mercantile or manufacturing districts it should not be necessary to travel more than 152 m (500 feet) to reach a fire alarm box and not more than 244 m (800 feet) in residential areas. Schools, hospitals, nursing homes, and places of public assembly should have a box near the main entrance.

The Municipal Grading Schedule does not require boxes in residential areas, but it does provide for a credit of up to 20 points if these areas are covered.

Radio   Radio frequencies and equipment should be provided so fire alarm operators are able to maintain communications with fire companies and essential fire department personnel away from their quarters. This permits more efficient operations including recall or reassignment of companies on the way to a response, reports from and between units on the fire grounds, and contact with units on in-service inspections and training.

To accomplish this, radio transmitter-receivers must be provided at communication centers, on apparatus in service, in chiefs' cars, on required reserve apparatus, and at all fire stations. Portable transmitter-receivers are required where there are 13 or more companies, but they are also very useful in smaller departments. When fire department frequencies are shared with another service or a nearby municipality, definite rules of procedure should be provided to avoid interference that hinders communications.

Fire Department Telephone Service   Surprisingly, telephones in homes and places of business are more widely used for reporting fires than are fire alarm systems. In some communities this is the only means, and as a result, the handling of telephone alarms is quite important.

There must be sufficient circuits from the commercial telephone system for receiving fire calls and they must be specially listed as such in the telephone directory. Telephone alarms from the public should be transmitted to a specific location, usually the communication center, in all cases, and not to any individual fire company. Private telephones in fire stations should not be listed in the directory. A device should be provided that automatically records all communications received by telephone and arranges for immediate playback.

All telephone as well as box alarms, if not transmitted automatically, must be retransmitted by a fire alarm operator to at least the first and second alarm companies over the required dispatch circuits. Multiple alarms should be retransmitted to all companies. Coded signals on outside sounding devices may be sufficient in departments having on-call or volunteer departments.

Fire Safety Control

While we discussed organization for fire prevention in detail in Chapter 4, we must concern ourselves here with its importance to a municipal fire defense system. The extensive use of hazardous materials and processes, the universal use of electrical equipment and wiring, and the importance of constructing buildings to withstand and prevent the spread of fire makes laws covering these subjects essential to good fire protection. *Enforcement is of equal importance* since lack of enforcement is equivalent to absence of laws. All too often, the administration of these three aspects of fire prevention is handled by three separate departments of city government: fire department, electrical department, and building department. Cooperation is essential if there is to be effective enforcement and smooth operation.

Hazardous Materials   This aspect of fire prevention is usually under the fire prevention bureau, preferably part of the fire department. Qualified members of the bureau should make frequent and regular inspections to recommend safeguards for special hazards. These should be supplemented by inspections by members of the fire department company in the area for ordinary hazardous conditions. Inspections should be made as often as required to properly enforce fire prevention regulations. In the more hazardous occupancies, this will generally require at least four inspections a year, preferably unexpected.

Flammable and compressed gases present a serious fire hazard in most communities. The adequacy of local laws or ordinances governing storage, handling and use of natural gas, liquified petroleum gas, oxygen, and other compressed gases must be judged by comparison to the Fire Prevention Code (as recommended by the American Insurance Association) or the Standards of the National Fire Protection Association. Fire prevention schedule ratings also consider the condition of installations having flammable or compressed gases in judging the quality of enforcement and the degree of hazard.

Similar attention is given to *flammable liquids* including inside and outside storage, transportation and use, fuel oil installations, dry cleaning plants, and application of flammable finishes. *Special hazards* considered include garages, combustible fibers, explosives, hazardous chemicals, dust and plastics, woodworking plants, metalworking processes, and radioactive materials. *Miscellaneous hazards* include incinerators, flammable fabrics, packing material, rubbish, and welding.

A local fire prevention bureau is also responsible for aspects of fire prevention other than the elimination of hazards. These other activities are considered in judging and rating the quality of such a bureau. They include seeing that obstructions to exits in places of public assembly are removed, finding and fixing defective automatic fire protection equipment, and reporting structures in need of repair that might create a hazard either to their occupants or to fire fighters responding to an alarm. A superior fire prevention program will include the inspection of dwellings on invitation and continuous public education.

Electricity   A city will normally have a fairly detailed set of regulations covering electrical wiring. These regulations are usually enforced by the electrical department's experienced and qualified electricians and inspectors. In considering local electrical laws, they are compared to the standards of the National Electrical Code, but a deficiency will be assessed in the absence of an agreement by the utilities companies not to furnish electricity until an installation has been inspected and approved. There should also be ordinances to provide for reinspections of old installations and the correction of defects that have developed, but these are rarely found.

Proper records should be kept of all licenses, permits, inspections, violations, and corrections.

Building Department   The details of a building's construction have a great deal to do with restricting the spread of fire. Many cities have elaborate building codes that cover every aspect of construction. An adequate set of regulations embodies the National Building Code, as recommended by the American Insurance Association. Enforcement, however, depends on having a building department with a qualified builder-engineer in charge and an adequate staff of trained assistants.

When rating the adequacy of local building laws, important features of the local code are compared to the standard, and enforcement is judged by conditions observed after inspection of buildings under construction. As with fire prevention laws, lack of enforcement of building laws is equivalent to absence of laws.

Features of the building code that are considered include permissible areas according to types of construction and protection; requirements for protection of exterior wall openings; vertical openings and communications through fire walls; and thickness or fire-resistance requirements for exterior and fire walls. Other items are requirements for fire-resistive construction; fire exits; parapets on outside or fire walls; and quality of material and workmanship. Other items that should be covered by a building code are heating, venting, and air conditioning, sprinkler equipment, and standpipe and hose systems.

Roof covering and allowable frame construction are also important factors in the spread of fire. The National Building Code requires all roof coverings to be at least Class C: effective against light fire exposures and having no flying brand hazard. This eliminates the use of wooden shingles or shakes except those that have been pressure treated, tested, and approved by Underwriters' Laboratories.

Building department records should include building applications, building permits, operations, and inspections, as well as plans of important buildings.

REVIEW QUESTIONS

1. What is the Grading Schedule for Municipal Fire Protection?

2. What are deficiency points and how are they used?

3. What are the four major categories of a fire defense system?

4. How is the percent of deficiency computed? How is that percentage converted into deficiency points?

5. How is the deficiency class determined?

6. How is the classification used by insurance companies?

7. What are the two principal features considered in grading a water system?

8. What is required fire flow and how is it determined?

9. Name three features of a water system that cause restriction of fire flow.

10. How is the required number of fire hydrants determined?

11. Name six reliability features of a water system that are considered in the Grading Schedule.

12. Name the five conditions that must be met for a fire department to be recognized.

13. What is the required number of fire department personnel on duty to obtain normal efficiency from an engine or ladder company?

14. What determines the number and location of engine and ladder companies?

15. Name some of the special equipment that might be required.

16. Name three operations at a fire that indicate an efficient fire department.

17. Name six conditions that adversely affect fire department operations.

18. Name two ways that communications are important to fire department operations.

19. Describe briefly how the following fire alarm systems operate:
    (a) Fire alarm telegraph system
    (b) Series telephone system
    (c) Parallel telephone system
    (d) Coded radio system

20. Name three categories in which laws and enforcement are important to municipal fire defense.

Glossary of Terms

Acceptance test Test conducted to determine whether new apparatus, especially pumpers, meet the specified pressure and discharge requirements.

Adapter A device for connecting hose couplings of the same nominal size but having threads of a different pitch or diameter.

Aerial A mechanically or hydraulically operated turntable ladder mounted on a ladder truck. The entire assembly with its auxiliary equipment is referred to as an *aerial*.

Aerial platform apparatus A mechanically raised platform mounted on fire apparatus and designed for rescue and fire fighting service.

Aircraft crash truck Specialized fire fighting apparatus designed to handle fires and accidents involving aircraft.

Air foam (mechanical foam) A type of foam for smothering flammable liquid fires. It is produced by adding a liquid foaming agent to water to make it capable of foaming in the presence of air introduced by mechanical action.

Apparatus Mobile vehicles that are used to transport fire department personnel, equipment, and appliances to fires or other incidents.

Apparatus floor The main room in a fire station in which fire apparatus is quartered.

Applicator A special pipe or nozzle attachment for applying foam (a "foam applicator") or water fog (a "fog applicator") to fires; usually identified by the size of hose used.

Archaeologist A scientist specializing in the study of early civilizations.

Arson The crime of intentionally setting fire to a building or other property.

Atmospheric pressure The weight of air or atmosphere. Normally at sea level this is approximately 100 kPa (14.5 psi).

Attack The actual physical fire fighting operation utilizing available personnel and equipment. The implementation of tactical plans on the fireground in an aggressive manner.

Automotive equipment Self-propelled equipment or apparatus.

Auxiliary Supplemental fire fighting equipment or personnel that may not be part of the regular first line fire fighting equipment or personnel; frequently it is makeshift or reconditioned equipment provided for emergencies.

Auxiliary pump A pump used on fire apparatus with a capacity rating below 1890 lpm (500 gpm) and used as a booster pump or high pressure pump.

Back pressure The pressure or energy required to overcome the weight of water due to elevation. For example, in fire department practice it is usual to add 34.5 kPa (5 psi) to the pump pressure for each floor of elevation that the nozzle is above the pump, to overcome back pressure caused by gravity.

Backup line Ordinarily a line of 2½ inch or larger hose laid in the event that the initial attack with small lines proves inadequate; sometimes, an additional hose line backing up the attack and protecting personnel using fog lines for close attack on a flammable liquid fire.

Beam The main structural member of a ladder supporting the rungs or rung block.

Bed ladder The lower section of an extension ladder into which the upper section or fly ladder retracts. The lower section of an aerial ladder that beds onto the truck frame.

Boat (fire boat) A boat with pumps, hose, monitor nozzles, and other fire fighting equipment and appliances and staffed as a company in the fire department.

Booster Small line equipment consisting of a water tank and pump using small hose lines and small nozzles.

Booster pump A pump of less than 1890 lpm (500 gpm) carried as an integral part of fire apparatus and used to supply streams through small hoses.

Cantilever position The position of an aerial ladder supported only at the base by the turntable and unsupported at the top. Recommended safe ladder loading is much less when the ladder is in this position than when it is supported at the top.

Cartridge-operated extinguisher This is a 2½ gallon extinguisher that uses a cylinder of carbon dioxide gas as the expelling agent. It is so arranged that when the extinguisher is turned over and dropped on its "head" a pin breaks the seal on the carbon dioxide cylinder and releases the gas.

Cellar pipe or **Cellar nozzle** A special nozzle used for attacking fires in basements or other below-floor or below-deck spaces.

Chain of command The formal structure of an organization, such that every member knows to whom he is responsible and to whom he must report; e.g., fire fighter to captain, captain to battalion chief, and so forth.

Chemical engine An apparatus used in fire departments in earlier years. It consisted of two chemical tanks mounted on a wagon and was used for delivering small streams.

Chemical foam A type of foam designed to extinguish flammable liquid fires. It is made by the chemical reaction of an alkaline salt solution and an acid salt solution to form a gas in the presence of a foaming agent, which causes the gas to be trapped in tough, fire-resistant bubbles.

Chief officer   An officer in the fire department with the rank of battalion chief or higher.

Class A pumper   The present performance standards for fire department Class A pumpers require that they deliver rated capacity at draft at 1034 kPa (150 psi) net pump pressure with a lift not exceeding 3.05 m (10 feet); they must deliver 70 percent of rated capacity at 1378 kPa (200 psi) and 50 percent rated capacity at 1723 kPa (250 psi).

Class B pumper   An earlier performance standard, adopted about 1912, required that pumpers deliver rated capacity at 827 kPa (120 psi) net pump pressure and that they deliver 50 percent capacity at 1378 kPa (200 psi) and 33⅓ percent capacity at 1723 kPa (250 psi).

Combination   A piece of fire apparatus designed to perform two or more functions.

Combustible   A material or structure that ignites and burns at temperatures ordinarily encountered at fires; a material that, when heated, gives off vapors that in the presence of oxygen (air) may be oxidized and consumed by fire.

Combustion   See *Fire.*

Company   A basic fire fighting organizational unit headed by an officer. A company is usually organized to crew certain types of apparatus, such as an engine company or a truck company.

Conflagration   Fire that extends over a large area, crossing natural or artificially created barriers in the process, and that destroys many buildings.

Consumption   In fire science, *consumption* usually refers to water used for purposes other than fire fighting; frequently called *domestic consumption.*

Convection currents   Currents of air that transfer heat from one place to another by circulation of heated particles of gas or liquid.

Curriculum   A course of study or a set of courses of study.

Deluge gun   A powerful stream device arranged to permit two or three 2½-inch hose lines to merge into one nozzle.

Deluge set   Deluge gun without the nozzle.

Distributing nozzle   A revolving nozzle with several vari-angular orifices.

Draft   To draw water from a static source into a pump that is above the level of the water supply. This is done by removing the air from the suction hose and pump and allowing atmospheric pressure to push water through the suction hose into the pump.

Elevating platform truck   See *Aerial platform apparatus.*

Engine company   A fire company equipped with a pumping engine. Sometimes called a "pumper company."

Excessive heat   Temperatures, generally in excess of 150°C. (300°F.), that tend to vaporize exposed fuels and enable fires to spread.

Exposure   Property that may be endangered by fire in another structure or by an outside fire. In general, burnable property within 12 m (40 feet) may be considered to involve an exposure hazard. The distance may be much greater in the case of very large fires, high winds, shingle roofs, or low humidity.

Extension ladder   A ladder of two or more sections that may be extended to various heights. The maximum height determines the name of the ladder, e.g., a 35-foot extension ladder, a 50-foot extension ladder, and so forth.

Extension of fire   Spread of fire, usually during the course of fire fighting operations, to areas not previously involved; an extension of fire through open partitions into an attic, or extension through openings to another room or building.

Extinguish   To quench; to put out flames; essentially, to completely control the fire so that no abnormal heat or smoke remains.

Fire   Rapid oxidation of combustible materials that results in light and heat.

Fire behavior   The manner in which fuel ignites, flame develops, and fire spreads. Sometimes used to distinguish characteristics of one particular fire from typical fire characteristics.

Fire confinement   That stage in fire fighting when there is no more possibility of fire extension.

Fire control   That stage in fire fighting when the fire is no longer spreading and eventual extinguishment is certain. In large forest or brush fires the term *fire containment* is sometimes used. (See also *Fire confinement*.)

Fire drill   In common usage, practice in evacuating buildings, or in other operations that might be necessary in case of fire. Also, a device used by primitive humans to start fires.

Fire fighting tactics   Methods of employing fire companies in an efficient, coordinated manner in the field so as to get satisfactory results.

Fire flow   The amount of water available or needed to extinguish a fire at a particular location.

Fire hazard   Any condition or thing that might cause or contribute to the danger of fire.

Firehouse   A fire department station housing fire department apparatus and fire department members while on duty. A fire station.

Fire mark   A metal plate fastened to a building in early days to designate the company that insured the building against fire.

Fire prevention   Any operation that tends to prevent fire from starting or spreading.

Fire protection   A term that includes fire prevention, fire control, fire extinguishment, fire detection, and fire investigation.

Fire pump   The main pump on a fire department pumper. A stationary pump approved for fire service, usually for private protection.

Fire resistance   A measure of the ability of a material to keep from burning.

Fire service   The organization that supplies fire prevention and fire fighting services to the community; its members, individually and collectively. Sometimes used in a broad sense to include all persons involved in fire protection.

Fire-stopped wall   A wall so constructed that the space between studs or other vertical members is blocked in order to avoid the vertical spread of fire.

Fire stream   A stream of water of suitable shape and velocity, directed from a nozzle and effective for the control of fires.

Fire technologist A person skilled in fire science, usually with two years of formal fire science education or equivalent experience.

Fire truck Any piece of motorized fire fighting apparatus.

First alarm company A fire company assigned to respond on the first alarm.

First company due The fire company expected to arrive first at the scene of the fire.

Flame spread The rate or speed at which flame spreads over the surface of a material.

Flow test Measuring the amount of water available for fire fighting from a water system at a particular location.

Fly The upper section of an extension or aerial ladder.

Foam A fluffy mixture formed by compounds introduced into a stream of water. Special nozzles or proportioning devices develop a stream of *tenacious foam* capable of smothering fires, especially those involving flammable liquids.

Foam blanket The covering of foam applied over a burning surface for a smothering effect.

Fog A jet or cloud of fine water spray discharged by special fog or spray nozzles.

Free burning Materials or structures lacking appreciable fire resistance and subject to quick hot fires. Also, the types of fires resulting from the burning of such materials in the presence of adequate oxygen.

Friction loss The loss of energy of water due to resistance to motion between the water and the insides of hose or pipe. This is the conversion of the water energy to unusable heat and it is therefore termed "loss."

Front Usually, the position in fire fighting assumed by the commanding officer; it is one of the first positions covered upon arrival of fire fighting units.

Front-mount pumps A fire pump mounted in front of the vehicle radiator; usually used in rural service.

Gate valve A control valve used to shut off or turn on water in a water system.

gpm Gallons per minute, a measure of the rate of water flow. The metric equivalent is lpm, liters per minute; one gpm equals 3.78 lpm. In fire fighting lpm and gpm are used to express the output of pumpers, hose streams, hydrants, and so forth.

Gravity tank An elevated storage water tank connected to a water system, public or private.

Guard control center The center of operations for a guard system in a large industrial plant to which all guards report; it is equipped for communication with various points inside the plant and emergency calls outside.

Heat Energy that is present at all temperatures above absolute zero. The student of fire science is primarily interested in heat above normal temperatures that is produced by burning or oxidation processes and that may cause exposed combustible material to burn.

Heat conductor Material capable of transmitting heat rapidly. Steel, for example, is a good heat conductor; heating one end of a steel bar causes the other end to become hot quite soon. Heating a piece of wood, on the other hand, has the reverse effect, and wood is therefore said to be a poor heat conductor.

Heat detectors Devices sensitive to rising temperatures or to a definite temperature and capable of sending an alarm, opening sprinklers, or causing any of several other operations necessary in case of fire.

Heavy streams Large caliber fire streams too heavy for manual operation of hoses and therefore applied through a powerful stream appliance.

Heel The base of a ladder. To take a position at the base of a ladder for raising or lowering is to heel a ladder.

High lift pump When water is raised to a higher elevation in steps, one pump delivering water from the original source to storage at higher levels, the pump taking suction from the higher storage is termed a *high lift pump*.

High pressure fog A small capacity spray jet produced at very high pressure and discharged through a small hose and gun-type nozzle.

High pressure pumper A pumper with special high pressure pumps and equipment used to supply small, high pressure fire streams. Also, a pumper purchased under specifications calling for a discharge of stated quantities at pressures higher than for normal pump performance.

Hooking up Connecting a pumper to a hydrant and connecting hose lines.

Hose A flexible tube or conduit used to convey liquids. Fire hose usually consists of a rubber tube covered with one or two woven jackets used to protect the inner liner.

Hose coupling Fire hose sections, usually in lengths of 15.25 m (50 feet), are connected together to produce a continuous hose line by means of hose couplings. Each length of hose has a male threaded shank on one end and a female swivel on the other end, with a gasket to make a leakproof connection.

Hose line One length or a series of connected lengths of hose; sometimes shortened to "hose" or "line."

Hose reel A reel permanently mounted on apparatus for small fire hose.

Hose wagon More commonly called *hose truck;* "wagon" is a carryover from the horse-drawn apparatus days. It is a fire truck chiefly used to carry hose and is frequently equipped with water tank and booster pump.

Hydrant A valved outlet on a water system with one or more threaded outlets, installed to provide fire department hose connections to supply water to pumpers or fire lines.

Hydraulically designed system Sprinkler systems designed to deliver the required water density, liters per square meter (gallons per square foot), to any place in the system.

Hydraulics A study of the use, movement, and properties of liquids. To the student of fire science hydraulics is particularly important in determining the action of water in pipe and hose lines.

Impeller The rotating part of a centrifugal pump that imparts energy to the water by means of centrifugal force.

Incendiary fire A fire started maliciously.

Incipient fire A fire of minor consequence or one in its initial stages.

Indirect application A technique of injecting water particles into the upper atmospheric level of a fire within a confined space. Its purpose is to generate steam and distribute unvaporized droplets of water, which then cool heated materials beyond the reach of a fire stream.

Indirect fire loss   Losses from results other than destruction by the fire itself; for example, loss of business because a plant or store is shut down as a result of a fire.

Initial attack   The first point of attack on a fire; the point where hose lines are used to prevent further extension of the fire and to safeguard life while additional lines are being placed in position.

Jacks   Ground jacks supporting the turntable of an aerial ladder; a lifting mechanism to raise heavy objects.

Knock down   To reduce flame and heat, usually by the use of hose lines, in order to prevent further extension of fire.

Ladder company   A fire company crewing a ladder or aerial truck and especially trained in ladder work, ventilation, rescue, forcible entry, and salvage work.

Ladder lock   A control for engaging the pawl of an aerial ladder. A control for securing a ladder in its bed when not in use.

Ladder pipe   A heavy stream appliance attached to an aerial ladder and capable of directing a large stream of water from a height.

Ladder truck   A fire department apparatus that carries ladders and other equipment.

Lay   A sequential method of stretching hose from fire apparatus.

Length   A 50-foot section of hose.

Lift   The height of the suction chamber of a pump above a static source of water.

Line   Usually refers to hose, but may refer to a length of rope such as a "life line."

Loss of life fires   Fires that cause death.

Low lift pump   A low lift pump takes suction from the original source and pumps water into storage at a higher level that is used as suction for another pump. (See also *High lift pump.*)

Male coupling   The threaded hose nipple that fits into the thread of a swivel coupling; a coupling to which nozzles and other appliances are attached.

Master stream   Any of a variety of heavy streams formed by utilizing large hose lines or merging two or more hose lines into a powerful stream device (see *Heavy streams*).

Midship pump   A fire pump located under or behind the driver's seat and supported on the pumper frame between the front and rear wheels of a pumper.

Monitor   A heavy stream device used to control large fire streams.

Mop-up   A late stage of fire fighting in which remaining hot spots are quenched, usually with small amounts of water.

Multiple-line underwriting   An insurance term designating services of insurance companies that provide a variety of kinds of insurance, for example, fire, casualty, and automobile.

National standard thread   A standard for the size and pitch of fire hose threads, established by the American National Standards Institute; today most fire hose threads meet this standard.

Nozzle   A device attached to a hose or appliance to restrict the area of flow, thus increasing the velocity of the water and forming a stream that will be effective in reaching and controlling the fire.

Nozzle operator   A fire fighter assigned to operate a nozzle on a hand line.

Nozzle pressure   More properly termed *velocity head,* this refers to the energy or force of the water when it comes out of the nozzle.

Nozzle reaction   When the water of a fire stream goes through a nozzle its velocity is rapidly increased, giving it additional force; the *nozzle reaction* is a natural physical force in the opposite direction that tends to push the hose and nozzle backward.

Nozzle tip   The head of the nozzle; interchangeability of tips easily allows altering the nozzle discharge.

Off-duty fire fighter   A paid fire fighter who is on personal time; i.e., is not available to respond to fires.

Officer in charge   The ranking officer in command at a fire.

Open up   To ventilate a building filled with smoke and heat so that hose streams may be advanced to extinguish the fire. Also, to enter forcibly into a closed burning building.

Outrigger jacks   Jacks designed to extend outward from the sides of an aerial or elevating platform truck to provide a wider base of support than that provided by the width of the truck chassis.

Overhauling   A late stage of the fire extinguishing process during which the areas involved are carefully scrutinized for any remaining trace of fire or embers. It also includes an effort to protect the property against further damage from the elements.

Oxygen   A gas present in the earth's atmosphere in about a 21 percent concentration that is an essential part of the fire triangle.

Oxygen deficiency   Insufficient oxygen to support life or combustion. This occurs when the oxygen content is about 16 percent or less of the atmosphere.

Pawl   The part of the mechanism of an extension ladder that secures the extensions to the rungs of the main ladder.

Pedestal   The operator's stand located on the turntable of an aerial ladder.

Performance test   See *Acceptance test.* Also, a test made to determine the condition of a pumper.

Plug   A fire hydrant.

Portable monitor   A monitor mounted on a stand that can be moved from one location to another.

Powerful stream appliance   An appliance used for handling heavy streams.

Prefire planning   Surveys of special hazards and plans for possible fire fighting operations.

Pressure   Certain kinds of force, such as the force water exerts against the inside walls of a pipe or hose. It is measured in kilopascals or pounds per square inch; one kPa equals 0.147 psi.

Pressure tank   A heavy steel tank partly filled with water and compressed air introduced to provide pressure; sometimes used instead of a gravity tank.

Private protection Fire protection provided by the owner or occupant of private property.

Probationary period A specified time during which a recruit in the fire department is on trial to determine whether the recruit is capable of becoming a qualified fire fighter.

Probationer One who is serving a probationary period.

Proportioner A device for introducing a predetermined amount of foam stabilizer or wetting agent into a fire stream.

Public protection Fire protection provided by a city, county, or other governmental body.

Pumper A fire truck equipped with a fire pump of at least 1895 lpm (500 gpm) capacity and carrying fire hose and other fire fighting equipment.

Pumper capacity The amount of water a pump is designed to deliver per minute at a pump pressure of 1034 kPa (150 psi).

Pumper ladder A combination apparatus that carries the standard equipment of both a pumper and a ladder truck.

Pump operator A fire department member whose duty it is to operate a pumper.

Quad A quadruple combination apparatus carrying (1) water tank, (2) hose, (3) pump, and (4) ladders. Sometimes called a quadruple combination truck.

Quarters The fire station to which a fire company or an individual is assigned.

Quint A combination fire apparatus having an aerial ladder in addition to the four standard equipment components of a quad.

Reaction See *Nozzle reaction.*

Relief valve A pressure-controlling device on a pump to prevent excessive pressure when a nozzle or hose line is shut off. Relief valves operate on the discharge side of the pump and when the pump pressure exceeds a predetermined amount the water bypasses the discharge, or a governor slows the speed of the motor to obtain the desired pressure. Relief valves are also used to relieve excessive pressures in pressure vessels.

Rescue The saving of life and removal of endangered persons to a place of safety.

Reserve Apparatus not in the first-line duty, but ready for relief duty in case of large fires, accidents, or breakdown of regular apparatus.

Residence requirements The requirements of the personnel or civil service department regarding where an applicant for the fire department must live.

Residual pressure The force that water exerts against the pipe or hose while the water is moving.

Respond To answer an alarm in accordance with a prearranged assignment or upon instruction of a dispatcher. To proceed to the scene of a fire or alarm.

Response The act of responding to an alarm. The entire complement of crew and apparatus assigned to an alarm.

Riser A vertical water pipe used to carry water for fire protection to elevations above grade, such as a standpipe riser or a sprinkler riser.

Rotary gear pump A positive displacement pump that employs closely fitting rotors or gears to force water through a pump chamber.

Rpm (rpm) Revolutions per minute used as a unit of rotating speed.

Rubber lined cotton hose A rubber or synthetic rubber tube protected by a woven cotton jacket or jackets.

Run Response to a fire or alarm. The term comes from the days of hand-drawn apparatus, when fire fighters ran to the fire pulling their apparatus.

Rung The round or step of a ladder.

Salvage Procedure to reduce incidental losses from smoke, water, fire, or weather during and following fires.

Salvage company An organized fire company or unit specializing in salvage operations.

Salvage cover A large waterproof canvas or other material used to cover contents of a burning building to avoid water damage.

Seat of fire Area where the main body of fire is located as determined by the outward movement of heat and gases.

Service aerial An aerial ladder mounted on a four-wheel truck; does not require a tillerman.

Service truck (city service truck) A ladder truck that carries only ground ladders and other truck equipment.

Shut-off nozzle A type of nozzle that permits the flow to be controlled by the operator.

Side wall sprinklers Sprinklers installed on the outside of a building to protect it from exposure fires.

Size-up The mental evaluation made by the officer in charge in order to determine a course of action.

Smoke A combination of gases, including carbon dioxide, and other products of combustion that hinders respiration, obscures visibility, and delays access to the seat of the fire.

Smoke jumper A fire fighter who jumps from an airplane and parachutes into the vicinity of fires, usually forest or brush fires.

Snorkle The trade name of an elevating platform truck manufactured by Snorkle Fire Equipment Co.

Spontaneous heating (combustion) A process that increases the temperature of a material without drawing heat from an outside source. If heating continues until the ignition temperature is reached, spontaneous ignition (combustion) occurs.

Spotting a ladder Positioning a ladder in reference to the objective to be reached.

Spray Water applied through specially designed orifices that form finely divided particles so as to more readily absorb heat and smother fire by the generation of steam.

Spray nozzle A nozzle constructed to produce a water spray; often called a *fog nozzle*.

Sprinkler head A term sometimes applied to the nozzles of an automatic sprinkler system. It usually consists of a frame holding a fusible link that covers an orifice. When heat opens the fusible link, water is forced through the orifice.

Standpipe A vertical water pipe riser used to supply water for fire fighting in buildings. It may also provide hose outlets for fire department use.

Static pressure The force exerted by the water in a system when the water is not moving.

Stationary pump A fire pump permanently installed to provide fire streams.

Steamer connection A siamese connection for pumping water into a sprinkler or standpipe system. Also applied to the large suction outlet of a hydrant.

Stoker The person who was assigned to shovel coal into a steam fire engine. Now obsolete.

Stored pressure extinguisher A fire extinguisher that is constantly under pressure due to compressed gas forced into it to act as an expellent for the extinguishing material.

Strategy The overall method and plan for controlling an outbreak of fire with the most advantageous use of available forces.

Suction A practice of taking water from a static source located below the level of the pump by exhausting air from the pump chamber and suction hose and using atmospheric pressure to push water into the pump.

Suction hose A hose reinforced against collapse, which permits its use for drafting.

Supervised circuit An electrical circuit that is designed to give a trouble signal in case the circuit is broken, grounded, or in any way becomes out of order.

Supervisory system A system such as a signaling or sprinkler system that is designed to give a specific signal if something goes wrong with the system or if for any reason it is not functioning.

Suppression The total operation of extinguishing a fire from the time of discovery.

Tactics Methods of employing fire companies in an efficient, coordinated manner in the field in order to get satisfactory results with the forces employed and to deny the fire any potential avenue of extension.

Telescopic boom A power-operated supporting apparatus for the platform of an elevating platform truck and constructed with retracting slider sections.

Tiller The rear steering wheel of a large tractor-tiller ladder truck or other vehicle requiring a separate steering wheel for the rear axle.

Tillerman The person who steers the rear axle.

Tips Nozzle tips that change the size of the orifice of a hose stream.

Transmitter-receiver A piece of radio equipment that enables the operator to either transmit or receive radio signals.

Triple combination A combination pumper carrying hose and water tank with booster pump.

Truck company A ladder truck company whose primary duties at fires are rescue, ventilation, forcible entry, and salvage.

Truss A structural member of a ladder, which joins and supports the ladder beams.

Turntable The rotating table or platform at the base of an aerial ladder that supports the ladder.

Turret   A heavy stream device permanently mounted on a pumper or fire boat.

Vacuum   A condition existing when much of the air and other gases have been removed from a contained area. Pressure less than atmospheric.

Vaporization   The physical change of going from a solid or liquid into a gaseous state.

Wagon   A hose truck, usually equipped with a water tank and booster pump or front mount pump.

Water curtain   A stream of water projected between a fire and an exposed structure to protect against extension of the fire.

Water mains   The pipes composing a water system. The term usually applies to the larger (4-inch or larger) pipes.

Water shed   A region of land that drains into a particular river or stream.

Water tower   A fire apparatus equipped with an extendable mast having one or more heavy stream nozzles. Used to apply master streams into upper floors of buildings.

Wetting agent   A chemical additive used to reduce the surface tension of water, thus providing greater penetration.

Wet water   Water to which a wetting agent has been added.

Appendix A

The Metric System

In this edition of the text we have expressed units of measure in both the metric system and the "English" system, the latter shown in parentheses. In 1973 when General Motors, the largest manufacturer in the country, announced it was switching to the metric system for all new products, it became evident that the United States as a nation was on the way to converting to this more logical system of weights and measures. Since that time many large industries have followed the same procedure and soon all of us, including the fire service, will be speaking in terms of centimeters instead of inches, meters instead of yards and feet, and kilometers instead of miles.

History   A decimal system of units was first conceived in the sixteenth century. Many years later in 1790 the French developed the metric system, which is a decimal system of weights and measures based on the meter from the Greek metron, a measure as the basic unit of length, and the gram as the basic unit of mass.

As international trade developed, it became apparent that there was a need for international standardization and for more accurately defined units. In 1872 therefore, there was held an international meeting in France attended by representatives of 26 countries to develop such standardization. As a result, the Treaty of the Meter was signed in Paris in 1875 by 17 countries including the United States. This established a permanent organization known as the General Conference of Weights and Measures, which was to meet every six years.

The United States became "officially" a metric nation in 1895, but the English system continued in general use. In 1975, however, President Ford signed the Metric Conversion Act of 1975, committing us to eventual total conversion.

In 1960, the international General Conference of Weights and Measures met and revised and modernized the metric system, and it is now known as SI, the International System of Units.

Advantages   The advantages of SI (Le System Internationale d' Unites) are many, but probably the most important is its internationality. A meter (or metre) is the same in any language or country, as are the gram, the liter (litre), and the watt.

Another advantage is that because it is a decimal system, to change from smaller to larger units it is only necessary to divide by multiples of 10, or to move the decimal point to the left.

The metric system has a logical development that is not found in the English system. The length of the meter (or metre) equals about one ten-millionth of the curved line between the North Pole and the Equator along the Greenwich Meridian (eventually established at 39.37 inches). Other units in the system are logically related to the meter. For example, the liter is the unit of volume and is equal to one cubic decimeter (a tenth of a meter, cubed) of fluid. The weight of a liter of water at its temperature of maximum density became the kilogram or 1000 grams, the basic unit of mass.

1960 Revisions   In 1960 when the International Conference made revisions to modernize the metric system and developed SI, it followed the same logical system. Most of the SI units are familiar, but there are two or three important to fire science that should be explained. The Joule (named for John Prescott Joule, 1818–1889, who formulated the laws of conversion of energy) has long been used as a measure of heat or work in the electrical field, but it is now used also to replace the calorie and the Btu as units of measure. The Joule is defined as the work done or the heat generated by a current of one ampere acting for one second through a resistance of one ohm.

Sir Isaac Newton (1642–1727) established the concepts of mass, force, gravity, and many other principles of mechanics. He may have contributed more to science than any other person in history, and it is appropriate, therefore, to give his name to the unit of force, the newton. The first law of motion is that the force is equal to mass times acceleration ($F = ma$), and the newton (N) is the force required to accelerate one kilogram of mass one meter per second, or $N = kg{\cdot}m/sec^2$.

The English system of units is confusing because it uses the pound to apply to both force and weight or mass. Pressure or stress is expressed in pounds per square inch (psi). Pressure or stress can be defined as the intensity of force, and it is expressed as a force acting upon a given area. In the SI the unit of pressure is called the Pascal (named for noted French mathematician and physicist Blaise Pascal, 1623–1662, and equals the energy of one newton acting on one square meter, or $Pa = N/m^2$. To convert pounds per square inch (psi) into kilopascals (kPa), multiply by 6.89.

Conversion　　In using the SI it is important to be familiar with the meaning of the prefix, if any, which denotes the value or magnitude of the unit, as well as with the unit's symbol. For instance, one *centi*meter equals one one-hundredth of a meter, and one *kilo*gram equals one thousand grams. The following table shows these prefixes, symbols, and values:

PREFIX	SYMBOL	VALUE	PREFIX	SYMBOL	VALUE
tera-	T	10^{12}	deci-	d	10^{-1} (0.1)
giga-	G	10^9	centi-	c	10^{-2} (0.01)
mega-	M	10^6	milli-	m	10^{-3} (0.001)
kilo-	k	10^3 (1000)	micro-	μ	10^{-6}
hecto-	h	10^2 (100)	nano-	n	10^{-9}
deca- (or deka-)	da	10	pico-	p	10^{-12}

The following tables show how to convert units most frequently used in fire science from the English system to the SI.

ENGLISH	MULTIPLY BY	TO OBTAIN SI	ABBREVIATION
AREA			
acres	0.4	hectares	ha
square feet	0.093	square meters	m²
square inches	0.0006	square meters	m²
square miles	2.6	square kilometers	km²
FLOW			
gallons per minute U.S. (gpm)	3.78	liters per minute	lpm
HEAT			
Btu (British thermal units)	2.33	Kilojoules per kilogram	kJ/kg
TEMPERATURE			
degrees F.	subtract 32 and then multiply by 0.556	degrees Celsius (Centigrade)*	°C.
LENGTH			
feet	0.305	meters	m
inches	2.54	centimeters	cm
miles	1.61	kilometers	km

MASS

pound	0.454	kilograms	kg
ton	0.907	kilograms	kg

POWER

horsepower	0.746	kilowatts	kW

PRESSURE OR STRESS

inches of mercury	3.39	kilopascals	kPa
pounds per square inch	6.89	kilopascals	kPa

VELOCITY

miles per hour	1.61	kilometers per hour	km/hr
feet per second	3.05	meters per second	m/sec

VOLUME

cubic feet	0.028	cubic meters	m³
U.S. gallons	3.8	liters	l

*One degree Celsius (formerly called Centigrade) equals one one-hundredth of the temperature change between the freezing point (0°C.) and the boiling point (100°C.) of water.

Appendix B

INDEPENDENT FIRE RATING ORGANIZATIONS

DISTRICT OF COLUMBIA
Insurance Rating Bureau of
  the District of Columbia

1101 15th St. N.W., Rm. 504
Washington, D.C. 20005

(202) 628-1230

HAWAII
Hawaii Insurance Rating
  Bureau

700 Bishop St., 5th Floor
P.O. Box 4500
Honolulu, Hawaii 96813

(806) 531-2771

IDAHO
Idaho Surveying and
  Rating Bureau, Inc.

1007 West Jefferson Street
P.O. Box 1069
Boise, Idaho 83701

(208) 343-5483

LOUISIANA
Property Insurance
  Association of Louisiana

Suite 1034, One Shell Square
P.O. Box 60730
New Orleans, Louisiana 70160

(504) 581-7972

MISSISSIPPI
Mississippi State Rating
  Bureau

2685 Insurance Center Drive
P.O. Box 5231
Jackson, Mississippi 39216

(601) 981-2915

NORTH CAROLINA
North Carolina Fire
  Insurance Rating Bureau

226 South Dawson Street
P.O. Box 2021
Raleigh, North Carolina
27602

(919) 834-4313

TEXAS
Texas Insurance Advisory   2801 So. Interregional Hwy.   (512) 444-9611
   Association   P.O. Box 15
                                  Austin, Texas 78782

WASHINGTON
Washington Insurance   801 Alaska Building   (206) 624-1822
   Examining Bureau, Inc.   P.O. Box 3967
                                  Seattle, Washington 98124

Washington Surveying and   1100 Alaska Building   (206) 622-8853
   Rating Bureau   P.O. Box 1168
                                  Seattle, Washington 98111

Effective March, 1976

Appendix C

Occupancy Hazard Classification

The text in several places refers to "light hazard occupancy," "ordinary hazard occupancy," and "extra hazard occupancy." An experienced fire protection engineer can classify the hazard of an occupancy by inspection, but the following guide, excerpted with permission from *Installation of Sprinkler Systems* (NFPA No. 13, copyright 1975, National Fire Protection Association, Boston), will help the novice recognize classes of occupancies.

Light Hazard Occupancies

Light Hazard—Occupancies or portions of other occupancies where the quantity and/or combustibility of contents is low and fires with relatively low rates of heat release are expected.

Light Hazard Occupancies include occupancies such as:

Churches	Museums
Clubs	Nursing or Convalescent Homes
Educational	Office, including Data Processing
Hospitals	Residential
Institutional	Restaurant seating areas
Libraries, except large stack rooms	Theaters and Auditoriums excluding stages and prosceniums

Ordinary Hazard Occupancies
Ordinary Hazard (Group 1)

Ordinary Hazard (Group 1)—Occupancies or portions of other occupancies where combustibility is low, quantity of combustibles is moderate, stock piles of combustibles do not exceed eight feet and fires with moderate rates of heat release are expected.

Ordinary Hazard Occupancies (Group 1) include occupancies such as:

Automobile Parking	Dairy Products Mfg.
Garages	and Processing
Bakeries	Electronic Plants
Beverage	Glass and Glass Products
Mfg.	Mfg.
Canneries	Laundries

Ordinary Hazard (Group 2):

Ordinary Hazard (Group 2)—Occupancies or portions of other occupancies where quantity and combustibility of contents is moderate, stock piles do not exceed 12 feet and fires with moderate rate of heat release are expected.

Ordinary Hazard Occupancies (Group 2) include occupancies such as:

Cereal Mills	Mercantiles
Chemical Plants—Ordinary	Machine Shops
Cold Storage Warehouses	Metal Working
Confectionery Products	Printing and Publishing
Distilleries	Textile Mfg.
Leather Goods Mfg.	Tobacco Products Mfg.
Libraries—large stack room areas	Wood Product Assembly

Ordinary Hazard (Group 3):

Ordinary Hazard (Group 3)—Occupancies or portions of other occupancies where quantity and/or combustibility of contents is high, and fires of high rate of heat release are expected.

Ordinary Hazard Occupancies (Group 3) include occupancies such as:

Exhibition Halls	Piers and Wharves
Feed Mills	Repair Garages
Paper and Pulp Mills	Tire Manufacturing
Paper Process Plants	Wood Machining

Warehouses (having moderate to higher combustibility of content, such as paper, household furniture, paint, general storage, whiskey, etc.)

Extra Hazard Occupancies

Extra Hazard—Occupancies or portions of other occupancies where quantity and combustibility of contents is very high, flammable liquids, dust, lint or other materials are present introducing the probability of rapidly developing fires with high rates of heat release.

Extra Hazard Occupancies include occupancies such as:

Aircraft Hangars	Explosives and Pyrotechnics
Chemical Works (Extra Hazard)	Woodworking with
Cotton Pickers and Opening	Flammable Finishing
Operations	

Under favorable conditions and subject to the approval of the authority having jurisdiction, a reduction of requirements to the next less restrictive occupancy classification may be applied to the following occupancies:

Cold Storage Warehouses	Machine Shops
Cotton Picker & Opening	Mercantiles
Operations	Metal Working
Feed Mills	Paper & Pulp Mills
Leather Goods Manufacturing	

Judgment must be applied in using this guide. For instance, hotels are listed under "Light Hazard Occupancies," but while the lobby and guest rooms are in the light hazard class, the kitchen and laundry areas should be considered as ordinary hazard.

Appendix D
Forms Related to
Fire Department Operation

PERSONNEL PERFORMANCE EVALUATION
DEPARTMENT OF FIRE

PART I

Type of Report _____

Period _____

NAME: _____ POSITION: _____ ASSIGNMENT: _____

PART II   PERFORMANCE OF DUTY FACTORS

RATERS		OUT STANDING	ABOVE AVER.	AVERAGE	BELOW AVER.	MARGINAL	UNSATIS-FACTORY
1ST LEVEL	2ND LEVEL	1	2	3	4	5*	6*

		ABILITY TO LEARN
		ADAPTABILITY: ADJUSTS TO SITUATIONS
		ALERTNESS: RESPONSIVE, PERCEPTIVE
		AMBITION: DESIRES IMPROVEMENT
		APPEARANCE: BEARING, NEAT, WELL GROOMED
		APPROACHABILITY: EASY TO DEAL WITH
		ATTITUDE: LOYALTY AND RESPECT FOR JOB
		COOLHEADEDNESS: DELIBERATE ACTION
		COMMON SENSE: LOGICAL AND PRACTICAL
		DECISIVENESS: CAN SELECT A COURSE OF ACTION
		DEPENDABILITY: ACCOMPLISHES DESIRED ACTIONS
		ENTHUSIASM: INTERESTED PARTICIPATION
		FIRE PREVENTION PERFORMANCE
		INGENUITY: ABILITY TO SOLVE PROBLEMS
		INITIATIVE: WORKS ON HIS OWN
		KNOWLEDGE OF FIRE SKILLS
		MAINTAINS PHYSICAL FITNESS
		MECHANICAL SKILLS
		ORAL EXPRESSION
		PUBLIC RELATIONS
		QUALITY OF WORK
		QUANTITY OF WORK
		SAFETY CONSCIOUSNESS
		TACT: AVOIDS UNNECESSARY OFFENSE
		WRITTEN EXPRESSION
		OTHER: _____

PART V   COMMENTS

CONTINUE COMMENTS ON REVERSE SIDE OF FORM

PART III   OVERALL VALUE

	OUT-STANDING	ABOVE AVER.	AVER.	BELOW AVER.	MARGINAL	UNSATIS-FACTORY
1ST LEVEL					*	*
2ND LEVEL					*	*

PART IV   PROMOTIONAL POTENTIAL

PART VI   SIGNATURES

RANK	ASSIGNMENT	DATE

1ST. LEVEL: _____

RANK	ASSIGNMENT	DATE

2ND. LEVEL: _____

RANK	ASSIGNMENT	DATE

REVIEWER: _____

I HAVE BEEN SHOWN THIS REPORT

MEMBER	DATE

* EXPLAIN IN COMMENTS

OFFICERS PERFORMANCE EVALUATION

Type of
Report: _____

DEPARTMENT OF FIRE

Period: _____

PART I

NAME: _____ POSITION: _____ ASSIGNMENT: _____

PART II PERFORMANCE OF DUTY FACTORS

RATERS		OUT STANDING	ABOVE AVER-	AVERAGE	BELOW AVER.	MARGINAL	UNSATIS- FACTORY
1ST LEVEL	2ND LEVEL	1	2	3	4	5*	6*

ACCEPTS AND ACTS UPON SUGGESTIONS
ACCEPTS RESPONSIBILITY FOR HIS ACTIONS
ADAPTABILITY: ADJUSTS TO SITUATIONS
ALERTNESS: RESPONSIVE, PERCEPTION
APPEARANCE: BEARING, NEAT, WELL GROOMED
APPROACHABILITY
ATTITUDE: LOYALTY AND RESPECT FOR JOB
COOLHEADEDNESS: DELIBERATE ACTIONS
DECISIVENESS: CAN SELECT A COURSE OF ACTION
DEPENDABILITY: ACCOMPLISHES DESIRED RESULTS
DEVELOPS SUBORDINATES
DISPLAYS PROFESSIONAL KNOWLEDGE
ENTHUSIASM: INTERESTED PARTICIPATION
EVALUATES SUBORDINATES ACCURATELY
EXERCISES PROPER DEGREE OF SUPERVISION
FIRE PREVENTION PERFORMANCE
FIRE TACTICS
FORCE: EXECUTES ACTIONS VIGOROUSLY
INGENUITY: ABILITY TO SOLVE PROBLEMS
INITIATIVE: WORKS ON HIS OWN
MAINTAINS HIGH STANDARDS
MAINTAINS PHYSICAL FITNESS
MERITS CONFIDENCE AND TRUST
ORAL EXPRESSION
PLANS FOR FUTURE NEEDS
QUALITY OF WORK
SAFETY CONSCIOUSNESS
SHOWS CONCERN FOR SUBORDINATES
TACT: AVOIDS UNNECESSARY OFFENSE
WRITTEN EXPRESSION

PART III OVERALL VALUE

	OUT- STANDING	ABOVE AVER.	AVERAGE	BELOW AVER.	MARGINAL	UNSATIS- FACTORY
1ST LEVEL					*	*
2ND LEVEL					*	*

PART IV PROMOTIONAL POTENTIAL

* EXPLAIN IN COMMENTS

PART V COMMENTS:

CONTINUE COMMENTS ON REVERSE SIDE OF FORM

PART VI SIGNATURES

RANK _____ ASSIGNMENT _____ DATE _____

1ST LEVEL: _____
RANK _____ ASSIGNMENT _____ DATE _____

2ND LEVEL: _____
RANK _____ ASSIGNMENT _____ DATE _____

REVIEWER: _____

MEMBER:
I HAVE BEEN SHOWN THIS REPORT   DATE _____

OCCUPANCY RECORD

ADDRESS		ZONE	DBA				

OCCUPANCY TYPE | MULT. OCC. | | RECOMMENDED INSPECTION FREQUENCY | INSP. | | SPRINKLERS
YES | NO | | RESP. | FULL | PART | B.O. | NONE

CONTACT | TITLE | PHONE | BLDG. TYPE(S) | ROOF COVERING

LAFD No.	NAME OF WELL	OWNER'S ADDRESS	WELL LOCATION

	YEAR ISSUED FOR	PERMIT FOR
DIV. 4 PERMITS ISSUED		
DIV. 5 PERMITS ISSUED	YEAR ISSUED FOR	PERMIT FOR
SPECIAL PERMITS ISSUED	YEAR ISSUED	PERMIT FOR

Date Inspected	INSPECTOR	ASSIGN-MENT	NOH OR SURVEY	F.N.	REMARKS	Date Completed

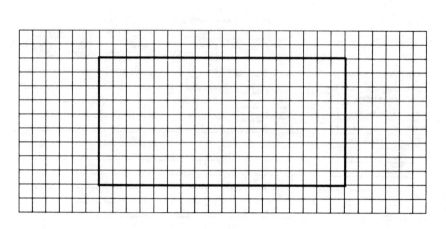

LEGENDS:
△	Access	E	Elevator	Ⓦ	Water
(AS)	Automatic Sprinkler	Ⓖ	Gas Shut-off	⫶	Stairway (Open)
Ⓔ	Electricity	≡HV≡	High Tension Wires		Other

To be used for diagraming small minor occupancies establishments for fire fighting inspections. Modify if necessary. Locate symbols in relation to the item. Use additional sheets if multiple storied buildings. Rectangular figure represents building.

OCCUPANCY RECORD
(FIREFIGHTING INSPECTION REPORT)

FIRE PREVENTION INSPECTION RECORD

I. Occupancy Address _____   Inspection Year 19_____

II. Building Owner Name   Business Address   Telephone

1st _____ _____ _____
2nd _____ _____ _____
3rd _____ _____ _____
4th _____ _____

| | | | Occ. | Cert. |
III. Name and Address of Responsible   Telephone   DBA   Class | U & O# |

1st _____ _____ _____ ___ ___

2nd _____ _____ _____ ___ ___

3rd _____ _____ _____ ___ ___

4th _____ _____ _____ ___ ___

IV. Name of Fire Department Inspector   Time   Date   Name of Business Representative

1st _____ _____ _____ _____
2nd _____ _____ _____ _____
3rd _____ _____ _____ _____
4th _____ _____ _____ _____

V. Inspection Details   REMARKS   Callback/ Spec.

	1	2	3	4		Date	Item	By
A. LIFE HAZARDS								
1. Exits and Aisles								
2. Occupant Loads								
3. Decorations								
4. Emergency Lighting								
B. FIRE PROTECTION								
1. Automatic Sprinklers								
2. Standpipes								
3. Fire Assemblies								
4. First Aid Appliances								
5. Fire Alarms								
6. Hydrants								
C. COMMON HAZARDS								
1. Housekeeping								
2. Heating Equipment								
3. Smoking								
4. Motors								
5. Electrical								
D. SPECIAL HAZARDS - PERMITS								
1. Vertical Openings								
2. Grease Ducts								
3. Flammable Liquids								
4. Other								
5. Other								
6. Other								

LEGEND:
   NA Not Applicable
   —   No Hazard
   ╱   Correction Needed (see remarks)
   ✗   Hazard or Violation Corrected

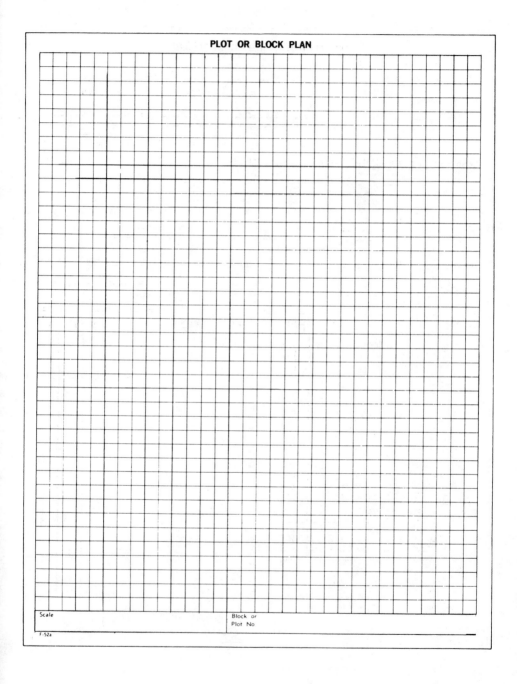

PLOT OR BLOCK PLAN

Scale

Block or
Plot No

F-52a

FIRE DEPARTMENT
PREPLANNING INSPECTION FORM

Street No. _____

	HOUSE KEEPING	INSPECTED	
		DATE	BY

ADDRESS _____ DATE _____

FIRM OR BLDG. _____

OWNER OF BLDG. _____ ADDRESS _____

BUSINESS MANAGER _____ HOME PHONE _____

RESPONSIBLES IN EMERGENCY

_____ HOME PHONE _____

_____ HOME PHONE _____

_____ HOME PHONE _____

OCCUPANCY _____ TYPE OF CONSTRUCTION _____

ROOF: TYPE _____ COVERING _____ FLOOR _____

HEIGHT _____ CEILING _____ INT. WALLS _____

SPRINKLERED _____ STANDPIPES _____

CONTROL BD. _____

FIRE DETECTION EQUIP: TYPE _____ MONITORED BY _____

HEATING SYSTEM: TYPE _____ CONTROLS LOCATION _____

AIR COND.:   TYPE _____ CONTROLS LOCATION _____

ELECT. SUPPLY: VOLTS _____ CONTROLS LOCATION _____

ATTIC ACCESS _____ AREA _____ SPRINKLERED? _____

BSMT ACCESS _____ AREA _____ SPRINKLERED? _____

GENERAL STORAGE AREA _____

FLAMMABLE LIQUID/HAZARDOUS STORAGE _____

HAZARDOUS PROCESS AREAS _____

VERTICAL OPENINGS _____ PROTECTED? _____

EXTERIOR EXPOSURES _____

MISCELLANEOUS/REMARKS _____

_____

_____

PREPLANNING INSPECTION FORM

All lettering, etc., shall read from the front of the building

– NORTH –

– WEST –

– EAST –

– SOUTH –

WATER SUPPLY: Size Main _____   **Nearest Hydrant** _____   Flow _____

Type of Hydrant _____

Vicinity Hydrants _____

SMALL UNIT OCCUPANCY RECORD

ADDRESS (INK)		ZIP CODE	BLOCK NUMBER

ITEMS BETWEEN BOLD LINES TO BE IN PENCIL

DBA	PERMIT REQUIRED YES ☐ NO ☐
OCCUPANCY	MULTIPLE OCCUPANCY YES ☐ NO ☐

CONTACT	TITLE	PHONE—BUSINESS — EMERGENCY
BUILDING OWNER ADDRESS		PHONE—BUSINESS — EMERGENCY

DATE	REMARKS (INK)	Entry By (Name)

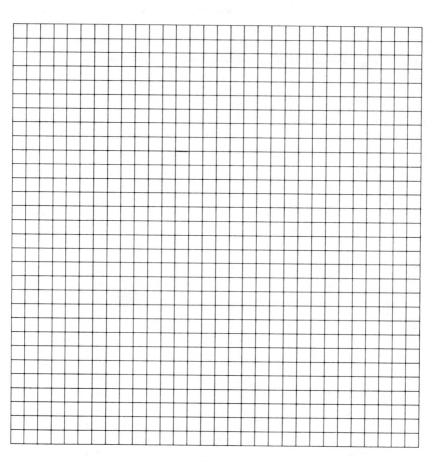

LEGENDS:

Ⓐ Access   ELV Elevator   Ⓦ Water

☩ Sprinkler Control   Ⓖ Gas Shut-off   IIII Stairway (Open)

Ⓔ Electricity   ≡ H V ≡ High Tension Wires   Other

DAY	**RESPONSE**	NIGHT	INSPECTION RECORD	
Engine		Engine	Date	Name & Assignment of Inspecting Member
Truck		Truck		
Other		Other		

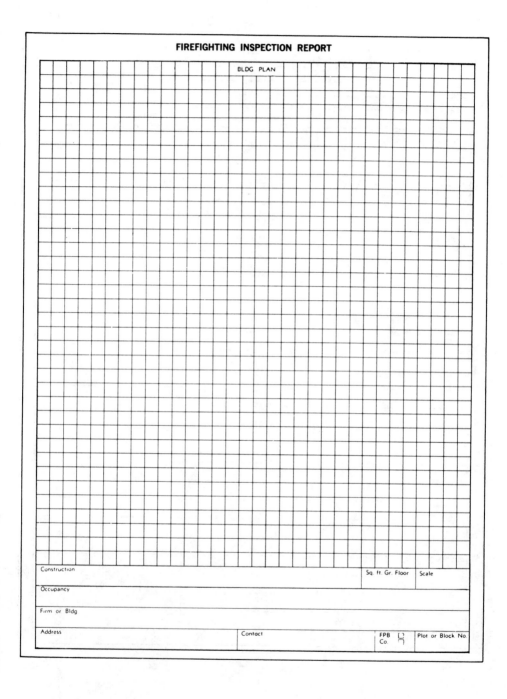

FIREFIGHTING INSPECTION REPORT

BLDG PLAN

Construction

Sq. ft. Gr. Floor

Scale

Occupancy

Firm or Bldg

Address

Contact

FPB Co.

Plot or Block No.

BATTALION CHIEFS MONTHLY REPORT

Month of_____

EMERGENCIES	This Month	To Date		This Month	To Date
No. alarms	_____	_____	Multiple alarms	_____	_____
No. responses	_____	_____	No. small lines	_____	_____
Running time	_____	_____	No. large lines	_____	_____
Pumping time	_____	_____	Water used	_____	_____

DISTRICT INSPECTIONS	This Month	To Date
A. Company fire prevention		
Routine	_____	_____
Night	_____	_____
Re-inspections	_____	_____
Referred to F. P. B.	_____	_____
Total F. P. inspections	_____	_____
B. Pre-fire planning	_____	_____
C. Incinerator	_____	_____
D. Hydrants		
Tested	_____	_____
Obstructions removed	_____	_____
E. Fire alarm box	_____	_____
F. Total district inspections	_____	_____

MISCELLANEOUS SERVICES	This Month	To Date		This Month	To Date		This Month	To Date
Incinerator permits - Sta. 1	_____	_____	Sta. 2	_____	_____	Sta. 3	_____	_____
Sta. 4	_____	_____	Sta. 5	_____	_____	Total	_____	_____
Daily burning permits-Sta. 1	_____	_____	Sta. 2	_____	_____	Sta. 3	_____	_____
Sta. 4	_____	_____	Sta. 5	_____	_____	Total	_____	_____
Bicycles licensed - Sta. 1	_____	_____	Sta. 2	_____	_____	Sta. 3	_____	_____
Sta. 4	_____	_____	Sta. 5	_____	_____	Total	_____	_____
Voters registered - Sta. 1	_____	_____	Sta. 2	_____	_____	Sta. 3	_____	_____
Sta. 4	_____	_____	Sta. 5	_____	_____	Total	_____	_____

Visitors to quarters Adults _____   Children _____ Total _____

Other_____

PROJECTS AND SPECIAL ACTIVITIES (Use reverse side of this form if necessary)

_____

_____

Report submitted by _____

MASTER FIRE APPARATUS RECORD

Equip. No._____ Type_____ Manufacturer_____

Manufacturer's No._____ Year_____ Purchase Price $_____

Cap. Impr. Proj. No._____ Date Accepted_____ Date In Service_____

Date in Reserve_____

Chassis
Model_____ Type_____ Number_____
Laden Wt.: Front Axle_____ lbs. Rear Axle_____ lbs. Gross_____ lbs.
Total Length_____ Width_____ Height_____ "
Turn Radius at 45°: Front Outer Wheel_____ ft._____ in. Rear Inner Wheel_____ ft._____ in.
Wheel Base_____ "   Angle of Approach_____° Angle of Departure_____°

Engine
Make_____ Type/Model_____ Cycles_____
Displacement_____ Booster Tank_____ gals   Compression Ratio_____
Fuel Tank_____ gals.   Rated_____ b.h.p. at _____ rpm governed speed

Pump
Make_____ Type/Model_____ Pump No._____ Capacity_____ gpm
Ratio_____ Suction Size_____ "   Class_____

YEARLY OUT OF SERVICE TIME
RECORD

Fiscal Year	Routine Maint.	Repairs	Accident	Fiscal Year	Routine Maint.	Repairs	Accident

MONTHLY VEHICLE REPAIR AND MAINTENANCE SUMMARY

Month of _____ 19____

	This Month	This Yr. To Date
Repairs by Fire Department	_____	_____
Repairs by Municipal Service Center	_____	_____

Apparatus Time O.O.S. for:

	This Month	This Yr. To Date
Routine Maintenance	_____ hrs.	_____ hrs.
Repairs	_____ hrs.	_____ hrs.
Most Time O.O.S. for One Unit	_____ hrs.	_____ hrs.
TOTAL OUT OF SERVICE TIME	_____ hrs.	_____ hrs.

Other pertinent information (include repairs caused by accidents):

_____

_____

_____

_____

_____

_____

_____
Apparatus Maintenance Clerk

MONTHLY HOSE REPORT SUMMARY

To: Fire Chief   Month of:_____

Hose Used	3"	2-1/2"	1-1/2"	1"F	3/4"-1"B	No. of Soft Suctions	Month Totals	Year To Date
Fires								
Drills								
Tested								
Repaired								
TOTAL FT.								

Hose Located

On Apparatus								
In Reverse								
TOTAL								

Remarks (Use Reverse Side If Necessary)

_____

_____

_____

_____

Distribution:
Original - Fire Chief
Copy - File   _____
                                     Name   Title

PUMPER TEST

Test Date: _____ 19___

____ Annual Service Test
____ Training
____ Other: _____

APPARATUS DATA

Pumper No. ____; Equip. No. ____; Manufacturer: _____; Model No. ____ Mfg. No. ____

ENGINE: Make: _____; Type-Model: _____; Cyls.: ____; Size & Stroke ____
Rated: ____ b.h.p. at ____ r.p.m. governed speed. Comp. ratio: ____ Eng. No. ____

PUMP: Make: _____; Type: ____; Pump No. _____; 250 lbs. ____
Gear Ratio: Engine to Pump: Cap'y.: ____ 200 lbs.: ____
Suction Size: ____ in.; Length: ____ ft.; with inner and outer strainers in place.

RATED CAPACITY: ____% Capacity @ ____ G.P.M. @ ____ PSI Net Pump Pressure.
____% Capacity @ ____ G.P.M. @ ____ PSI Net Pump Pressure.
____% Capacity @ ____ G.P.M. @ ____ PSI Net Pump Pressure.

TEST SUMMARY INFORMATION

Location of test: _____ Elevation: ____ ft.
Operator: _____ Attested by: _____

Time Min.	R.P.M. Engine	R.P.M. Pump	Tip Size	N.P.	Disch. G.P.M.	Pump Press.	Vac. Lbs.	Net Press.	Dry Vacuum Test.
1st Test									Dropped ____ in.
2nd Test									in ____ minutes.
3rd Test									

Ign. – Battery Test: ____; Quick Lift Test: ____
Excess Power Test: Engine R.P.M.: ____; Pump Pressure: ____
with ____ inch nozzle tip.. Pitot Press.: ____; Vacuum ____ in

RECOMMENDATIONS:

W O R K   S H E E T

Time	DISCHARGE SIDE					SUCTION		ENGINE				REMARKS
	Pump Pressure PSI	Pitot #1		Pitot #2		Lift Ft.	Vac. In. Hg.	Eng. Tach.	Hand Counter	Oil Press.	Water Temp.	(Ignition-Perform., Etc.)
		PSI	GPM	PSI	GPM							

TEST NO. 1: CAPACITY   Hose and Nozzle Layout: _____

Road Gear:
Pump Stage:

Pressure Control Set at _____ lbs.; Pump Operating at _____ lbs.; Shut-Off Pressure Increase at _____ lbs.

TEST NO. 2: 200 LBS.   Hose and Nozzle Layout: _____

Road Gear:
Pump Stage:

Pressure Control Set at _____ lbs.; Pump Operating at _____ lbs.; Shut-Off Pressure Increase at _____ lbs.

TEST NO. 3: 250 LBS.   Hose and Nozzle Layout: _____

Road Gear:
Pump Stage:

Pressure Control Set at _____ lbs.; Pump Operating at _____ lbs.; Shut-Off Pressure Increase at _____ lbs.

LADDER TEST FORM

Reason for test: ___ Annual ___ Other   Test date: _____ 19____
Other remarks: _____

Equipment No. _____; Ladder No. _____; Usable length of ladder:_____ Ft.

Type Ladder:
    ___ Ground   ___ Aerial   ___ Elevating Platform   ___ Other
    ___ Folding   ___ Roofing   ___ Straight   ___ Extension

Manufacturer:_____

Material:_____

Date of purchase: _____19___; Date of last test:_____ 19____

	PASS	FAIL
TEST NO. 1 (Visual Inspection)	__	__

    Remarks:_____

TEST NO. 2 (Recovery Test)   __ __

    a. Measurement - no load applied: _____ inches
    b. Measurement - load applied: _____ inches
    c. Measurement - load removed: _____ inches
    d. Difference between a & c: _____ inches
    e. Test load applied: _____ pounds

TEST NO. 3 (Rung Twist Test)   __ __

    Remarks:_____
    Test weight applied: _____ pounds

TEST NO. 4 (Rung Stress Test)   __ __

    a. Measurement - no load applied: _____ inches
    b. Measurement - load applied: _____ inches
    c. Measurement - load removed: _____ inches
    d. Difference between a & c: _____ inches
    e. Test load applied: _____ pounds

TEST NO. 5 (Aerial Ladder Test)   __ __

    a. Measurement - rung to ground: _____ inches
    b. Test Load ____lbs. applied:
    c. Measurement - rung to ground: _____ inches
    d. Difference between a & c: _____ inches
    Remarks: _____

RECOMMENDATION:
_____
_____
_____

Tested by: _____

APPARATUS MAINTENANCE GUIDE & RECORD

Month of_____ Equip. No#_____ Designation_____ Station_____

CHANGE OF PLATOON (Check the following immediately after roll call)

Glass	Wiper Blades	Brakes (Service and Parking)
All lights	Position of switches	Pump control and valves
Water level (tank)	Radiator water level	Gas and oil level
Priming oil level	Breathing equipment	Mirrors
Proper location of equipment	Seat belts	

Date   ____ ____ ____ ____ ____
Platoon on duty   ____ ____ ____ ____ ____

FRIDAY - WEEKLY (Check and initial)

Check emergency warning equipment   ____ ____ ____ ____ ____
Check and clean battery   ____ ____ ____ ____ ____
Check tires/air pressure   ____ ____ ____ ____ ____
Check and clean compartments   ____ ____ ____ ____ ____
Check all nozzles, wyes, etc.   ____ ____ ____ ____ ____
Check all valves, etc.   ____ ____ ____ ____ ____
Check governor/relief valves   ____ ____ ____ ____ ____
Check portable pumps   ____ ____ ____ ____ ____
Check portable generators   ____ ____ ____ ____ ____
Clean motor and undercarriage   ____ ____ ____ ____ ____
Drain moisture from air tanks (brakes) ____ ____ ____ ____ ____
Check back up alarm   ____ ____ ____ ____ ____

MONTHLY - FIRST FRIDAY OF REPORTING PERIOD (Check/clean the following)

Extinguishers - Inventory - Salvage Covers - Drain & refill tank - Operate pump - Check & clean tools - First aid equipment - Perform dry pump test - Check for hose change - Check aerial & snorkel (hydraulic system, slides, cables)

Date_____ Checked by_____

REPAIRS AND PARTS REPLACEMENTS			Time O.O.S.	
Date	Nature	Where repaired/replaced	Routine Maint.	Repairs
____	_____	_____	___	___
____	_____	_____	___	___
____	_____	_____	___	___
____	_____	_____	___	___

SERVICE NEEDED/REMARKS (Use reverse side if necessary)

_____

_____

                                        Company Officer_____

INDIVIDUAL FIRE ALARM BOX RECORD

LOCATION _____ MASTER TO _____ BOX NO. _____

DATE INSTALLED _____ TYPE _____ MAKE _____ MOUNTING _____ CIRCUIT _____

BOX LIGHTED _____

	OPERATED			FEATURES TESTED				CONDITION					
Date	Test	Fire	False	Shunt	Non-Inter Success	Ground Return		Box	Pole	Decal	Light	Remarks	By

MONTHLY FIRE ALARM REPORT

Month of _____

Item	This Month	This Year	To Date Last Year	Total Last Year
Fire Alarm Boxes				
Added	___	___	___	___
Removed	___	___	___	___
In–Service	___	___	___	___
Activated for.				
Fire	___	___	___	___
False	___	___	___	___
Tested	___	___	___	___
Painted	___	___	___	___
Total Circuit Footage				
Overhead Open Wire	___	___	___	___
Overhead Cable	___	___	___	___
Underground Cable	___	___	___	___

Circuits Not Receiving Alarms This Month
(Circle ones applicable)   1   2   3   4   5   6   7

Tested Box _____ Circuit _____ Tested Box _____ Circuit _____ Tested Box _____ Circuit _____
Tested Box _____ Circuit _____ Tested Box _____ Circuit _____ Tested Box _____ Circuit _____

Circuit Interruptions

Date	Time	Circuit	How Detected	Action Taken	Date Completed
___	___	___	___	___	___
___	___	___	___	___	___
___	___	___	___	___	___

Abnormal or Defective Circuit Conditions

_____
_____
_____
_____
_____

(Use reverse side, if necessary)

	Date	Duration	Condition
Emergency Power Generator Tested	___	___	___
Battery Operation Test	___	___	___

_____
Fire Alarm Record Coordinator

COMPANY HOSE LOCATION RECORD

Station _____   Company _____   Equip. No. _____   Date _____   Co. Officer _____

$2\frac{1}{2}$" Hose

Left Bed   Right Bed

$1\frac{1}{2}$" Bundles

Top   Bottom

1" F or B Hose

$1\frac{1}{2}$" Live Lines

Top   Bottom

3" Hose

Spare Hose

3" $2\frac{1}{2}$" $1\frac{1}{2}$"

INDIVIDUAL FIRE HOSE RECORD

Length
I.D. No._____ Size _____ Type of Jacket _____ Type of Coupling _____

Date of
Purchase_____ Mfr. _____ Vendor _____ Cost/ Ft. $_____

Date Placed
In Service _____ Assigned to_____ Date_____ Reassigned _____ Date _____

| | Hose Record Test | | | | Hose Repair Record | |
Date	Condition	Remarks	Date	Type	Performed By
1.					
2.					
3.					
4.					
5.					
6.					
7.					
8.					
9.					
10.					
11.					
12.					
13.					
14.					
15.					
16.					
17.					
18.					
19.					
20.					

Date Hose Condemned_____

Reason _____

_____

Remarks:_____

_____

_____

_____

HOSE CONDITION CODE

"S" - Satisfactory
"D" - Damaged

   1 - coupling
   2 - acid burn
   3 - thermal burn
   4 - cut
   5 - mildew
   6 - other (explain in remarks)

COMPLAINT RECORD

Complaint No. _____
Received by   _____
Date   _____

Information Received:

Complainant _____
Address   _____ Telephone _____

Nature of Complaint _____
_____

Location of Complaint _____

Initial Action:

Assigned to _____ Date _____

Investigation date _____ Complaint Valid _____ Not Valid _____

Description of Hazard _____
_____

Notice of Violation: Date issued _____ Mailed _____ Delivered _____

Owner ( ) Tenant ( ) : Name _____
Address _____ Telephone _____

Reinspection date _____ Remarks _____
_____

Follow-up Action:

Date _____ Compliance: yes ( ) No ( ) Extension granted : yes ( ) no ( )
If yes, reinspection date _____ Second N.O.V. issued: date _____
Remarks _____
_____

Second Reinspection: date _____ Compliance: yes ( ) no ( )
Remarks _____
_____

Other Action:

Notice of Hearing: date _____ Time _____
Results _____

Reinspection after Hearing: date _____ Compliance: yes ( ) no ( )
Remarks _____
_____

Final Disposition:

Referred to: Attorney's Office _____ Public Works _____ Bldg. Dept. _____
             Other _____ Date _____

Remarks _____
_____
List any additional calls or referrals relative to this record

Appendix E
Inspection Guides

GENERAL INFORMATION

- Legal name of owner
- Name of building
- Location of building
- Type of building (office, warehouse, store, etc.)
- Name of resident agent or manager and telephone number
- Is there ready access to building by the fire department?
- Number of persons occupying the building
    by day
    at night
    greatest number at any time

MERCANTILE BUILDINGS

- Year constructed
- Date(s) of addition
- Dimensions, frontage, depth, area
- Height, in stories and feet, with or without basement
- Roof construction: metal, composition, gravel, asbestos composition, wood shingle, slate, tile, tar paper.

Parapets: width, height, condition

Chimneys: foundations, masonry, supports, joints and bends, timber supports, cracks

Walls: brick, concrete, reinforced concrete, cinder block, concrete block, steel, aluminum sheet, fiber glass, galvanized sheets, brick veneer, tile, stucco, tar paper, etc.

Floors: earth, wood, masonry, concrete, plastic

Note: Checklist questions appearing under theaters, hotels, schools, and dwellings merely supplement the general list carried under Mercantile Buildings; they do not replace them.

Foundation: open, wood, piers or columns, enclosed by masonry or concrete

Lighting System: condition good, poor, dangerous

Heating system: steam, hot water, gas, stoves, space heaters, radiant panels
- Placement of heating units
- Protection of heating units
- Clearance between parts of system and combustible materials?
- Clearance between parts of system and combustible walls?
- Condition of connections, flues, etc.

Boiler or Furnace Rooms:
- Separated from rest of building?
- In protected, noncombustible room?
- Secondary emergency exit?
- Proper electrical control system?
- Self-closing fire-rated doors?
- Adequate air intake system?
- Emergency shutoff controls for furnace?
- Emergency gas shutoff?
- Type of fuel used and method of fuel storage
- Method of disposal of ashes

Incinerators and Rubbish Rooms:
- Incinerator room cut off by fire-resistive walls and ceilings rated at not less than one hour?
- Equipment well away from stored combustibles?
- Flue protected by spark arrestors of proper gauge?
- Does the incinerator room have self-closing fire doors?
- Adequate outside intake for fresh air?

Flammable Liquid storage:
- Does bulk storage of flammable liquids or liquified petroleum gases meet code standards?
- Inside storage of flammable liquids within permissible limits?
- Proper "No Smoking" and "Danger" signs posted where flammable liquids are stored, dispensed, or used?

Housekeeping, Storage, and Waste Disposal

- Accumulated waste, rubbish, old furniture, paints, or other combustibles in or around the building?

- Waste removed from building daily?

- Kind of containers used to collect and store waste paper and other combustibles

- All cleaning materials noncombustible?

- Any flammable liquids stored on the premises, including duplicator fluids and thinner for cement in office areas?

Air Conditioning and Ventilation Systems

- Location of power room. Show on plan.

- Type of refrigerant used, if any

- Shutoff for refrigerant location? Show on plan.

- Combustibles located near air intakes?

- Interconnection of areas by ductwork or system? Show on plan.

- False ceilings to hide ductwork?

- Ductwork termination points?

- Fire-resistive cutoff in ducts at each floor level?

- Fusible link fire dampeners in ducts at all fire walls?

- Fire dampeners at intervals throughout duct length?

- Smoke detection alarm system and emergency automatic shutoff in ductwork system?

Kitchens, Serving Pantries, and Coffee Shop

- Cut off from the rest of the building by fire-resistive walls and ceilings?

- Self-closing fire doors between kitchen and dining room it serves?

- Serving counters protected by self-closing fire doors?

- Kitchen or pantry protected by an automatic sprinkler system?

- Does kitchen have range hoods? If so, height from floor, distance from combustibles and condition of filter system.

- Wiring under hood in conduit?

- Automatic shutoff for gas to cooking equipment, activated by extinguishing system?

- Shutoff system for deep-fat fryers?

- Carbon dioxide or dry powder extinguishers in sufficient number in kitchen area?
- Access doors to kitchen ductwork?
- Cleaning records for kitchen ductwork?
- Automatic sprinkler system for kitchen ductwork?

Electrical Equipment:
- Overheating?
- Panels, boxes, motors, and systems closed and secured against dust, foreign matter, and unauthorized access?
- Standby power system for exit lights, exit signs, and fire alarm system available and in operating condition?
- Properly fused circuits?
- Multi-outlet devices in use?
- All outlets carrying more than 120 volts plainly marked?
- Use and condition of extension cords?
- Condition of power cords to all electrical appliances?

Interior Construction of Building
- Enclosed or open shafts and stairways; describe.
- Fire cutoff construction between floors?
- Hatchways and vertical openings between floors? List.
- Presence, absence and condition of fire doors
- Fire cutoff construction in partitions and walls?
- Exposed joists, studs, and woodwork?

Construction and Finish of corridors and Hallways:
- Dead end corridors?
- Combustible wall or ceiling materials?
- Free passage allowed or barred?
- Smoke barriers in all corridors with doors in operating condition?
- Wall finish in good repair?

Stairways:
- Condition: good, poor, dangerous
- Constructed of fire-resistive material?

- Equipped with automatic fire doors?
- If without sprinkler protection, in fire-resistive enclosure?
- Obstructed by or used as storage space for any type of material?

Exits:
- All exit doors open outward?
- All exit doors equipped with panic hardware or without locks?
- Can exit doors be locked in any way that will prevent opening them from the inside?
- One exit sign for every emergency escape route and exit?
- Exit signs always lit?
- At least two exits from each corridor and hallway?
- At least two exits from the basement?
- Does exit path lead through more than one room?

Fire escapes:
- Condition: good, poor, dangerous.
- Construction meets code standards?
- Windows within five feet have wired glass?
- Windows or doors that open on a fire escape have wired glass?
- Obstructions on fire escape?
- Route blocked by parking or trash storage?
- Fire escape kept free of ice and snow?

Furnishings and Decoration
- Curtains and drapes of flameproof material?
- Fabric articles, used in chemistry laboratories, home economics, fine arts and industrial arts classes, flame-resistive or flame proof?
- Certificates of fire proof treatment of textiles or other materials?
- Kinds of carpeting in use in rooms, offices and hallways.

Fire Protection Systems

Fire Alarm
- Does the building have a fire alarm system? Check one:
  - ☐ completely automatic
  - ☐ manually operated
  - ☐ manually operated boxes on an electric circuit

☐ push button or switch on electric current

☐ handbell

- Do all occupants know how to use the system?
- Alarm connected to a sprinkler system and triggered when the system activates?
- Does an automatic fire detection system, heat activated or smoke activated, sound the alarm?
- Alarm stations on every floor of the building?
- Alarm stations within 200 feet of any place in the building?
- Alarm device or bell distinctive?
- Can the alarm bell be heard in all sections of the building?
- Fire alarm connected ahead of the distribution panel for electrical current?
- When was the system last tested?
- Direct connection to the fire department alarm system?

Fire Extinguishers

- One extinguisher for every 100 feet of corridor, lobby, or stairway?
- Extinguishers plainly marked or identified?
- Extinguishers in all kitchens, heater and furnace rooms, large storerooms and shop areas?
- Extinguishers readily accessible?
- Nozzles in good condition?
- Hoses securely attached and in good condition?
- Do extinguishers need recharging?
- Dates of inspection

Sprinkler Systems

- Complete or partial system?
- If partial, does it protect the following? Check those covered:

☐ paths of exit

☐ basements, cellars, boiler rooms, and spaces not occupied by people

☐ storerooms

☐ areas where combustible materials are used

☐ both sides of unprotected openings that would let smoke or flame spread and make an area untenable

- Sprinkler heads in good condition?
- Piping in good condition?
- All valves work and all supply valves are in open position?
- Adequate water supply? Source?
- Water pressure in the system?
- Date of last test
- Outside siamese connection to the system readily accessible?
- Does siamese connection have standard threads?
- Outside connection plainly marked and visible from street or road-way?
- Type of sprinkler system used: check one:
  ☐ regular
  ☐ deluge
  ☐ dry pipe
  ☐ carbon dioxide
  ☐ foam

Standpipe Systems

- Does the standpipe system extend to the top of the building?
- Adequate water supply? Source?
- Water pressure at the following heights: 2nd, 4th, 6th, 8th, 15th, 22nd, 30th stories, and at five story intervals above this?
- Hose cabinet on each floor?
- Kept locked and clean?
- Size hose and nozzle in cabinets?
- Condition of hose
- Date of last inspection and replacement.
- Hose nozzles screwed on tightly and free from obstruction?
- Main shutoff valves in open position?
- Do the threads on the standpipe outlets within the building match those on fire department hose?
- Do the threads of the standpipe connection match fire department hose threads?
- Outside connection visible from street?
- If not, sign showing where it is?

Occupancy Hazards
  • What is the nature of the business?
  • If manufacturing, what is produced?
  • If processing, list principal materials involved
  • How many employees are there?
  • Size and contents of packing room, if any
  • Materials used in packing
  • Size and contents of storage room (if more than one, list each separately).
    Show type of storage: piles, bags, boxes, crates, cans, on shelves, etc.
  • Open flame devices in use? check those found:
    ☐ welding torch
    ☐ cutting torch
    ☐ bunsen burners
    ☐ process furnaces
    ☐ drying oven
  • Any oils on the premises? Check type found:
    • Vegetable, animal, or mineral?
    • Amounts of any oil found?
    • How stored?
  • Explosives used or stored on the premises?
  • Oxidizing agents used in any process?
  • Combustible metals used in any process?
  • Condition of equipment: good, fair, poor, dangerous
  • Flammable liquids used or stored? List those found.
  • Dust allowed to accumulate on places other than the floor?
  • Evidence of poor or careless safety practices? Describe.

Exposure Hazards
  • Give the distances to all exposed buildings.
  • Height of each exposed structure?
  • Type of construction of each exposed building?
  • Kind of business conducted in each exposed building?
  • Are any exposed buildings protected by a blank wall and, if so, what
    type of construction?
  • Do any exposed buildings have parapets? If so, give heights and con-
    dition.

• Do any exposed buildings have wall openings protected by steel shutters?

• Which, if any, exposed buildings have wall openings protected by wired glass?

• Which, if any, exposed buildings are protected by a water curtain sprinkler system?

• Are any protected by an outside deluge system?

• Are there any avenues for fire travel? If so, specify material and nature of route.

THEATERS (Use applicable parts of check list for buildings plus the following:)

• Walls and ceiling of noncombustible construction?

• Joints tight enought to prevent smoke discharge?

• Asbestos fire curtain between stage and audience?

• Projection room equipped with automatic shutters on all openings in case of fire?

• Nothing in projection booth but film in use and required amount of film cement?

• Film scraps or ends stored in metal containers?

• Exhaust fan to remove smoke and fumes from projection room?

• Film rewound in projection booth?

• If rewound elsewhere, is area protected?

• Film stored in protective metal cabinet or closed metal shipping containers?

• Smoking prohibited in projection booth or where film is handled?

• Sand or water pails provided for disposal of carbon rods from arc lights?

• Smoking permitted only in protected and marked areas?

• Dressing rooms kept clean and free of trash?

• Space, if any, between ceiling and roof supports?

• Material used in roof supports?

• Flies constructed of wood or metal?

• Flies controlled by rope or metal cable?

• Trapdoors or shaftways on or backstage?

• Distance between stage lights and combustible materials adequate?

HOTELS (Use building checklist where applicable plus the following:)

- Extent of connection between rooms by air conditioning or ventilating ducts
- Combustible materials stored near air-intake ducts?
- Fan shutoffs? Show location on hotel plan.
- Kind of refrigerant used in air conditioning system
- Kind of refrigerant used in food storage cabinets
- Location of shutoffs for refrigerant gases or liquids
- Furniture repair shop within hotel?
- Carpenter shop within hotel?
- Plumbing and electrical repair shops within hotel?
- Cleaning and tailer shops located within hotel?
- Laundry located within hotel?
- All sources of flame or heat within any of above properly protected against fire spread? If not specify where lacking.

SCHOOLS (Use applicable parts of checklist for large buildings plus the following:)

- Two separate exits from all assembly rooms?
- At least two exits from each floor?
- Fire drills regularly scheduled and held?
- Time to evacuate building during drill?
- Drill observed by inspector? Date
- Staff properly trained in fire drill procedure?
- Furnace room separate from school building?
- If not separate, located in protected noncombustible room?
- Flammable liquids kept in metal containers with flame-arrestor spouts?
- Bulk storage of flammable liquids in drums with pump dispensers and return drip, located outside building?
- Quantity storage of flammable liquids in metal cabinets with flame-arrester vents?
- Gasoline engine powered devices stored inside school?
- Waste paper collected and stored in metal cans?
- Hot plates in offices or teacher's lounge?

- Smoking regulations posted and enforced?
- Number, spacing, and location of fire extinguishers adequate?
- Dates of inspections of extinguishers

Dwelling houses (Use building check list where applicable:)
- Condition of roof
- Composition of roof
- Chimmney(s) on solid foundation?
- Chimmney(s) cracked, or with open joints or loose bricks?
- Chimmney in good condition at all points of contact with interior, including roof?
- Combustible waste materials in yard? Check those found:
  dry grass, weeds, branches, paper, boards, other
- Garage housekeeping?
- Sheds well kept?
- Flammable liquid storage?
- Quantities of flammable liquids in house?
- Accumulated waste or discarded furniture in cellar?
- Condition of furnace or other heating source: good, poor, dangerous.
- Condition of flues and ducts: good, poor, dangerous.
- Furnace, stove, or smoke pipe too close to combustible ceiling or walls?
- Space heater improperly insulated and protected?
- Automatic shutoff on gas-fired appliances?
- Gas lines tight and leak free?
- Oil burner installation sound?
- Cleanout door at base of chimney?
- Workshop area kept clean?
- Paints, varnish, oil, and turpentine stored properly?
- Fire stops at bottom of wall studs in basement?
- Location of basement exits
- Electrical system in good repair?
- Type of fuse box
- Is there a 220 volt circuit?
- Hazardous extension cords?
- Multiple plugs in outlets or fixtures?

			INSPECTION GUIDE
			HOTEL AND APARTMENT
NO.	**ITEM**	**CODE**	**REQUIREMENT**
1.	ELECTRICAL	20.16 ELECT CODE*	Maintain lighting circuit fuses at 20 amps. or less. Flexible cord over 12 feet must be approved for hard usage. Check for temporary and illegal wiring. Maintain clearance around electrical panel.
2.	FLAMMABLE LIQUID	C.A.C. DIV 30	Storage shall be prohibited except that which is required for maintenance and operation of building and equipment. Such storage shall be kept in closed approved containers stored in a storage cabinet or in safety cans or in an inside storage room not having a door that opens into that portion of the building used by the public.
3.	HEATING & COOKING EQUIPMENT	C.A.C. 1.35 DIV 20 VENT CODE* PLUMB CODE*	Approved portable gas heaters (U.L. etc.) connected by rigid pipe or semi-regid metal connector equipment with shut-off valve. Check for proper ventilation as applicable (intake and exhaust). Keep combustibles away from heaters. Discourage the use of unvented gas heaters (during use of unvented heaters, adequate ventilation must be provided). Check furnace room for building separation, ventilation, combustible storage and location of fuel shut-off. Water heater area free of combustible storage, mops, rags, wax and polish containers.
4.	EXIT, HALLWAY, STAIRWAY, FIRE ESCAPE	C.A.C. 1.35 DIV 10 BLDG CODE*	Exits, hallway, stairways, and fire escape shall remain free of all obstructions. Required exits shall open from the inside within 30 seconds without the use of a key, tool, implement, or special knowledge. Penthouse door may be equipped with a moveable bolt or by a lock which can be opened from the inside without the use of a key, tool, or implement. Exits identified with sign (5″ letters) and directional arrows as necessary. Hallway and stairway adequately illuminated. Condition of fire escape.

*For final action see Company Standards.

5.	ATTIC, BASEMENT, ROOF	C.A.C. BLDG CODE* DIV 10 20.16 DIV 21	Seal all openings in wall, ceiling and draft stops made for electricity, plumbing, and vent installation that may extend the spread of fire into concealed spaces. Maintain antennas, electric wire and other obstructions at least 7 feet above roof (except legal guy wire). Roof storage prohibited. Basement storage prohibited unless sprinklered.
6.	FIRE PROTECTION APPLIANCES	C.A.C. BLDG CODE* 1.35 21.21 DIV 130 DIV 140	One 2A fire extinguisher within 75 feet of travel. This does not apply to single story nor to multiple story occupancy having no interior hallway. Carports and garages (4 cars) one 2A fire extinguisher within 75 feet of travel. Fire extinguisher serviced yearly, tagged, hung with top not over 5 feet above floor in accessible location. Check interior standpipe fire hose, nozzle, hose rack. Exterior standpipe valves, gaskets, caps, sprinkler system (18″ clearance) identifying signs (3″ letters). Fire warning system and appliance accessibility.
7.	FIRE SEPARATIONS	C.A.C. BLDG CODE* 1.35 20.16	Penthouse fire door equipped with (U.L.) self-closing device required. Check boiler room. Building and occupancy separation (fire doors). Recommend transoms above doors be removed and permanently sealed. Rubbish and laundry chute self-closing shutter, free of obstructions.
8.	COMBUSTIBLE STORAGE	C.A.C. DIV 10 DIV 21	Prohibited on roofs, attic, sub-floor levels, space under stairway and in any area not designated for human occupancy. Maintenance areas adjacent to exits free of combustible storage.
9.	COMBUSTIBLE DECORATION	C.A.C. 20.16 22.01	All drapes, curtains, etc., in hallway, lobby or other public areas shall be non-combustible or flame retardant treated by a state licensed applicator and properly tagged.
10.	COMBUSTIBLE REFUSE	C.A.C. BLDG CODE* 20.16 DIV 21	Adequate container, safely stored in an area set aside for such storage while awaiting disposal. Storage within building, non-combustible container and one-hour storage room.
11.	FIRE PERMIT	4.03	Required for hotel, apartment hotel or motel (3 or more stories and 20 or more guest rooms). Apartment house does not require a fire permit. When required, refer to F.P.B. Public Safety.

| 12. | EMERGENCY PROCEDURE | 10.40 20.01 | Recommend procedure for fire emergency be developed. Evacuation of building and Fire Dept. notification (which is required). Know how to call Fire Dept. and use fire protection appliances. Maintain conspicuous address. Accessibility to premises and unobstructed access to main electric and gas service shut-off. |
| 13. | RESPONSIBLE RESIDENT REQUIRED, SIGN | C.A.C. 20.17 130.02 | Responsible resident in charge required on premises. 16 apartment and hotel with 12 guest rooms. Fire regulation sign required in apartment house and each unit of establishment used for transients. |

INSPECTION GUIDE

ASSEMBLAGE OCCUPANCY

NO.	ITEM	CODE	REQUIREMENT
1.	AISLES	57.10.04 BLDG. CODE 91.33.11 TITLE 19, ART. 33	Provide at least one 44" wide aisle leading to each required exit.
2.	CANDLE DEVICE	57.05.10 A3 57.110.15	General approval required before use in city.
3.	CAPACITY	BLDG. CODE, DIV. 33 TITLE 19, ART. 33	Room capacity determined by Building Department. Fire Department may reduce for cause. 'Pink' to building department for capacity sign.
4.	DECORATIONS	57.22.01	Combustible decorations which constitute a fire hazard must be flameproofed. See Definitions (57.02) for 'Flameproofed.'

5.	EXITS	BLDG. CODE, DIV. 33 TITLE 19, ART. 33 DIV. 10, FIRE CODE	Basic exit and hardware requirements in Building Code. For exit obstruction corrections use Fire Code. Immediate corrections required.
6.	EXTINGUISHERS	57.140.10	One 2-A extinguisher within 75' of travel is required. In addition a minimum 4-BC provided adjacent to cooking facilities.
7.	KITCHENS	57.10.04 57.20.30 57.20.31	If exit is provided through kitchen, a 44" aisle should be maintained. Grease accumulations in exhaust systems are prohibited.
8.	OVERCROWDING	57.10.30	Number of persons allowed is limited to legal capacity.
9.	PERMIT	57.04.03E 57.02	A Fire Permit is required where there is a capacity of 50 persons or more. Refer to FPB for processnig. See Definitions of Assemblage Occupancy.
10.	RUBBISH	DIV. 21 & 47.20.16	Proper containers. Disposed of each day.
11.	SMOKING	57.110.02	Smoking not permitted in theaters and little theaters. No smoking allowed on any stage. See definition of 'Theater' and 'Little Theater.'
12.	OPEN FLAME	57.110.15	Allowed only under Special Permit. This includes fire dancers, tiki torches and similar uses.
13.	COMPRESSED GASES	CHIEF REG. #2	Allowed only under Special Permit. LFG cooking prohibited in buildings.

Index